ALLEN COUNTY PUBLIC LIBRARY

P9-EDO-037

3-23-77

CELESTIAL

HORIZONS

CELESTIAL HORIZONS

A Concise View of the Universe

John C. Rosemergy

Allyn and Bacon, Inc.
Boston · London · Sydney · Toronto

Cover photos from Yerkes Observatory.

Copyright © 1977 by Allyn and Bacon, Inc.,
470 Atlantic Avenue, Boston, Massachusetts 02210.

All rights reserved. Printed in the United States of America. No part of the material protected by this copyright notice may be reproduced or utilized in any form or by any means, electronic or mechanical, including photocopying, recording, or by any information storage and retrieval system, without written permission from the copyright owner.

Library of Congress Cataloging in Publication Data

Rosemergy, John C   1921-
  Celestial horizons.

  Includes bibliographies and index.
  1. Astronomy.   I. Title.
QB45.R69   523   76-26874
ISBN 0-205-05571-0

1963789

Contents

3 The Universe: The Scope and Pattern of the Cosmos

Motion and Gravitation: Celestial Mechanics 4

Radiation: The Stream of Energy and Information from Space 5

6 Stellar Evolution: Birth and Death in the Cosmos

Preface

When I heard the learn'd astronomer,
When the proofs, the figures, were ranged
   in columns before me,
When I was shown the charts and diagrams,
   to add, divide, and measure them,
When I sitting heard the astronomer where he
   lectured with much applause in the lecture-room,
How soon unaccountable I became tired, and sick,
Till rising and gliding out I wandered off by myself,
In the mystical moist night-air, and from time to time,
Look'd up in perfect silence at the stars.

                                        Walt Whitman

Those of us who attempt to write for or instruct beginning astronomy students—particularly students who may not be science-prone—can perhaps receive some guidance from the above lines from Walt Whitman.

I have tried to make this book different from most other astronomy texts in the following ways:

1. The book is short (and thus inexpensive), concise, and straightforward—not encyclopedic. It is an extremely flexible text, designed for a one-quarter or one-semester course in astronomy or for a unit of astronomy within a course in earth science or physical science. Rather than requiring the instructor to consume class time elucidating lengthy content, the book provides a basic foundation upon which the instructor can readily build to suit a particular class. The book provides for a variety of course activities—lecture, laboratory work, outside reading, etc.

2. Only those topics that seem to be most important to an understanding of elementary astronomy have been selected. Chapters 1–3 develop a view of the cosmos, beginning with our solar system and concluding with the expanding universe of galaxies. Chapters 4–6 consider physical principles (particularly of

force, motion, radiation, and nuclear changes) which explain astronomical phenomena and relate astronomy to the physical sciences.

3. The book begins where I believe the study of astronomy naturally begins, with naked-eye observation, and ends where such study should lead, in contemplation of unanswered and disturbing questions.

4. The book's content is tied together with a historical thread.

5. Suggested activities at the end of each chapter can be done at home with readily available materials.

I hope that the following features will also be helpful to the instructor and student:

1. The suggested readings at the end of each chapter are current (except where timeless), readable, and, in most cases, easily located.

2. Topics and questions listed at the end of each chapter do more than assist the student to review his or her study; they can stimulate class discussion and lead to outside reading.

I am grateful to the following astronomers and teachers who reviewed the manuscript and provided helpful suggestions: Irene M. Engle, Donald A. Farnelli, Richard C. Hall, Freeman D. Miller, Oren C. Mohler, William P. Beres, William H. Tauck, and George Reed. I am also indebted to many individuals and organizations from whom I obtained photographs. I am particularly grateful to my wife, Margaret L. Rosemergy; this book would not have been completed without her assistance and forbearance.

Ann Arbor, Michigan

John C. Rosemergy

CELESTIAL

HORIZONS

FIGURE
1.1
A photograph of the northern sky centered on Polaris, the
North Star. How long was it exposed? (Lick Observatory
photograph.)

THE TURNING SKY

1

The Space Age arrived on October 4, 1957, with the orbiting of Sputnik I, but this date marked the beginning of only another era in the history of a very ancient science. People have always looked into space and wondered what or who was out there. The story of human efforts to understand the design of the universe and our place in it is one of the most significant and fascinating aspects of intellectual history.

Astronomy has helped to solve some very earthly problems. Positions of heavenly objects have been charted for centuries in connection with timekeeping and navigation. And the search for knowledge of the cosmos has resulted in a clearer understanding of fundamental principles of physics and chemistry.

The study of astronomy can enhance one's appreciation of the beauty of the universe. The wheeling panorama of the heavens is one of the loveliest aspects of our environment. Most people who live in the Northern Hemisphere need only lift their eyes to enjoy the soft beauty of Cygnus on a summer evening or the sparkling brightness of Sirius and the majesty of Orion on a crisp winter night. Watching the inexorable arrival and departure of the constellations in season can provide a sense of orderliness and certainty amidst transitory affairs. Much in the record of human thought demonstrates that contemplation of the universe can enrich the quality of life.

Intermingled with the development of astronomy has been the development of the pseudoscience of astrology, which attributes much of human personality, behavior, and fate to pervasive cosmic influences. Astronomy owes a great deal—some even say its birth—to astrology. The import of cosmic influences is obvious; our very lives are sustained by the energy of the sun. But how pervasive are those influences? Do the heavens decree that you will do well or poorly on your next examination? Is such a question flippant, or is it profoundly searching?

Our study of astronomy should begin where the science itself began—with *observation* of the sky. The first portion of this chapter summarizes observations that can be made by simply looking up at the sky for a few hours over a period of a few weeks. If possible, you should make the observations described in the suggested activities on pp. 43–45 before beginning to study this chapter. The chapter will be more meaningful if you have observed the situations it describes.

CELESTIAL MOTIONS

The Turning Celestial Sphere

What does one see of the heavens with the unaided eye? Viewed from the earth, the universe appears to be a star-studded sphere with the earth at its center. The spherical sky seems to turn constantly around a stationary earth. The axis of the celestial sphere is tilted so that the top of the sphere is not directly overhead. The top is in the northern part of the sky and is marked by the North Star (assuming that we are viewing the sky from the Northern Hemisphere of the earth). The North Star appears to be nearly motionless, since it marks the approximate top of the turning sphere. The bottom of the celestial sphere is opposite the North Star—out of sight beneath the southern horizon.

As the sky turns, certain celestial objects rise on the eastern half of the horizon and set on the western half. Celestial objects in the northern part of the sky appear to circle the motionless North Star in a counterclockwise direction. The apparent motion of objects rising and setting in the southern sky is clockwise. (See Figures 1.1 and 1.2.)

The sky seems to turn a little more than once in 24 hours. The "little more" than one turn is not readily noticed, but in a year it adds up to one more apparent turn of the sky. It follows then that the "extra" turning of the sky in one month is about equal to the amount of turning in two hours; on a given night each star will set two hours earlier than it did a month before. If you note the position of a star at 10:00 p.m. tonight, you will see it in approximately the same position at 8:00 p.m. one month from tonight. (See Figures 1.3 and 1.4.)

FIGURE
1.2A

The apparent motion of setting stars. (Lick Observatory
photograph.)

FIGURE
1.2B

The apparent motion of stars near the eastern horizon.
What is the observer's latitude? (Wellesley College photograph.)

FIGURE
1.3

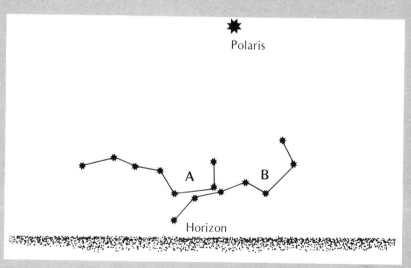

The sky's "extra turn." If the Big Dipper in Ursa Major is at
position A at 8:00 p.m. on December 15, it will be at position B at 10:00 p.m. on December 15; it will appear to have
moved 30 degrees on an arc centered at Polaris, the North
Star. At 8:00 p.m. on January 15, it will be at position B.

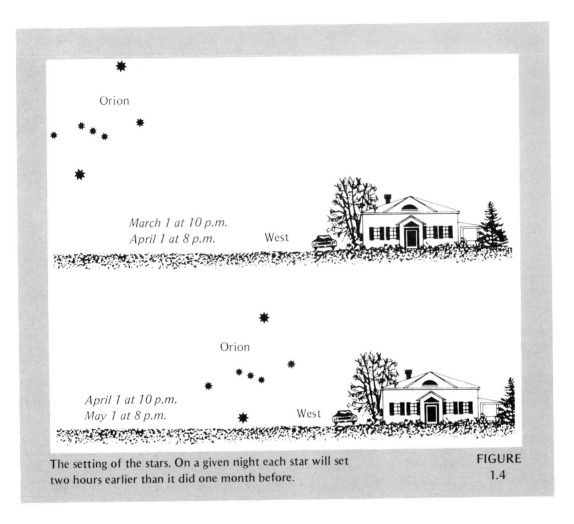

The setting of the stars. On a given night each star will set two hours earlier than it did one month before.

FIGURE 1.4

As the sky turns, the stars always remain in the same relative positions. The constellations and patterns of stars remain intact. The Big Dipper, for instance, retains its shape as it circles the North Star.

The Wanderers

All of the phenomena described above may be observed easily. Less easily noted are the motions of the seven "wanderers"—the sun, the moon, Mercury, Venus, Mars, Jupiter, and Saturn. The word *planet* is from the Greek for *wanderer*. These seven objects rise and set with the turning of the celestial sphere, just as the stars do; however, they do not seem to be fixed to the celestial sphere. Although they are carried along with the turning sphere, they appear to drift among the fixed stars. Mercury and Venus are sometimes visible in the eastern sky before sunrise; they are sometimes seen in the western sky after sunset. The maximum separation of Venus and the sun is considerably greater than that of Mercury and the sun. Venus can rise as much as three hours ahead of the sun and set that much later than the sun. Mercury's rising and setting times are never much more than 90 minutes different from those of the sun.

Mars, Jupiter, Saturn, the sun, and the moon all drift eastward against the background of the stars along the same general path. They have two principal apparent motions: (1) they rise and set with the turning of the celestial sphere, and (2) they seem to move around the earth in an easterly direction. The moon makes its trip around the earth in about a month (exhibiting its changing phases all the while). The sun appears to make an eastward circle around the earth in a year. The times for Mars, Jupiter, and Saturn are approximately 2 years, 12 years, and 30 years, respectively.

Some of these apparent eastward motions are much more obvious than others because of their comparative swiftness. The motion of the moon can be seen easily. It is interesting to watch the moon move in front of a bright star and occult (hide) it. The sun's apparent eastward motion is also readily evident, even though the sun and the stars can not be seen simultaneously. This apparent motion can be recognized by considering the position of the sun among the stars just after sunset when the stars first come into view. Stars that are very low in the west at sunset one day will already be below the horizon at sunset a few days later and hence no longer visible in the evening sky. As already noted, stars will set tonight about two hours earlier than they did a month ago.

It is the apparent eastward motion of the sun that accounts for the seasonal panorama of constellations. Stars that are high in the sky at midnight tonight will be there (but unseen) at noon six months from now. In six months the sun will appear to have moved halfway around the earth in its eastward circuit of the heavens. The sun at noon six months from now will be in front of stars that are in the midnight sky now.

The apparent eastward motions of Jupiter and Saturn are very slow, as already noted. Mars appears to drift more quickly; considerable change of position is evident in a period of a few weeks.

Early observers were immensely puzzled by the periodic interruptions that can be seen in the eastward drifting of Mars, Jupiter, and Saturn. These planets will sometimes seem to move westward for a while against the starry background and then resume their eastward drift. The westward motion is called *retrograde motion*. (See Figure 1.5.)

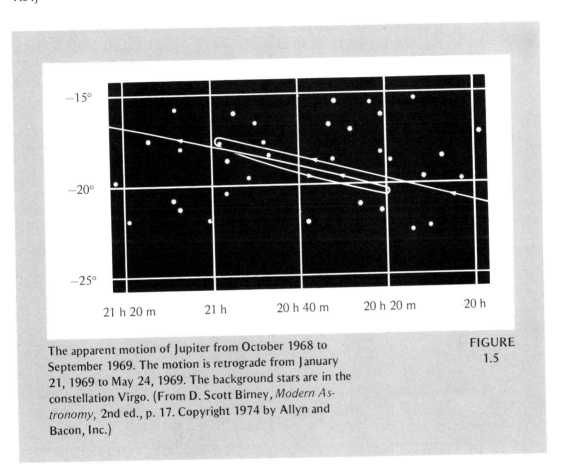

The apparent motion of Jupiter from October 1968 to September 1969. The motion is retrograde from January 21, 1969 to May 24, 1969. The background stars are in the constellation Virgo. (From D. Scott Birney, *Modern Astronomy*, 2nd ed., p. 17. Copyright 1974 by Allyn and Bacon, Inc.)

FIGURE
1.5

Phases of the Moon

The changing appearance of the moon is a particularly striking phenomenon associated with its monthly circuit across the starry background. The moon regularly passes through a cycle of phases: new moon, waxing crescent, first quarter, waxing gibbous, full moon, waning gibbous, last quarter, waning crescent, new moon. A gibbous moon is more than half-full (quarter) but less than full. The new moon is unseen, for it rises with the sun, moves across the sky with it, and sets with it. A night or two after new moon, a narrow crescent of the waxing moon may be seen setting on the western horizon soon after the sun sets. On each of the next few evenings the crescent grows larger and sets later. The *cusps* (points) of the crescent are always turned away from the sun. About one week after new moon, the first-quarter moon is in the south at sunset. During the next week a waxing gibbous moon is in the southeast at sunset. About two weeks after new moon, the full moon rises at sunset and sets at sunrise. During the third week after

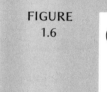

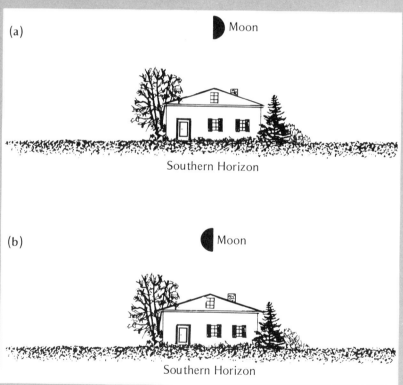

The phases of the moon. In which phase is the moon in diagram (a)? In diagram (b)? Approximately what time of the day or night would the moon appear as it does in diagram (a)? In diagram (b)?

new moon, the waning gibbous moon rises after sunset and can be seen in the southwest after dawn. During the last week before the next new moon, the waning crescent moon is in the southeast before sunrise. (See Figure 1.6.)

EFFORTS TO EXPLAIN CELESTIAL MOTION

How can the foregoing observations be explained? The earliest efforts to explain such observations were made long ago, for astronomy is an ancient science. The high quality of observation and thought of some of the early observers often surprises the student of today. However, some of the early work was a mixture of fact and fancy; some whimsical hypotheses have been offered concerning the design of the universe, the earth's place in it, and the motions in the sky. For example, ancient Hindus pictured a flat earth supported by four pillars; the pillars were supported by elephants standing on the back of a giant turtle. The most ancient attempts to explain the observations described in the first part of the chapter will not be discussed. Rather, attention will be directed to the two obvious possibilities for hypotheses on celestial motions.

There are two basic ways to account for the apparent celestial motions—either the sky moves or the earth moves. The earth *appears* to be stationary within a turning celestial sphere. "Common sense" caused this to be the central premise of the generally accepted hypothesis until only 400 years ago, although as early as 2,000 years ago a few scholars were questioning the prevailing theory. In 1543 Copernicus stated with uneasiness that he believed the earth moved; in an attempt to make his statements seem less new and startling, he quoted from Plutarch (46?–120?):

> The rest hold the Earth to be stationary, but Philolaus the Pythagorean says that she moves around the (central) fire on an oblique circle like the Sun and Moon. Heraclides of Pontus and Ecphantus the Pythagorean also make the Earth move, not indeed through space but by rotating round her own centre as a wheel on an axle from West to East.

The Ptolemaic System

It appeared obvious to most ancients that a great star-studded sphere turned constantly around a stationary earth. However, there was no obvious explanation for the wanderings of the five visible planets, the sun, and the moon. The generally accepted explanation for these wandering motions was devised by the Greek Hipparchus in the second century B.C. His work was refined and extended by Claudius Ptolemy in the second century A.D. Ptolemy was a Greek mathematician, astronomer, and geographer who worked at Alexandria. His classic book is known by the Arabic name *Almagest*, meaning "the greatest of books."

Ptolemy continued to picture the earth as being at the center of the universe and the fixed stars as being on a heavenly sphere, the Primum Mobile, which turned in 24 hours. Between the earth and the celestial sphere, the moon, Mercury, Venus, the sun, Mars, Jupiter, and Saturn moved counterclockwise about the earth in nearly concentric orbits. In addition to their independent motions around the earth, these "wanderers" shared in the daily motion of the entire heavens imparted by the Primum Mobile; this accounted for their daily rising and setting. The centers of the orbits of these seven "wanderers" did not coincide perfectly with the center of the universe; the centers of the orbits were located somewhat off-center at points termed *equants*. The five planets among the seven "wanderers" did not simply move counterclockwise in circular orbits. Each was considered to be moving in a smaller circle which moved on the larger circle. The large circle traced by the moving center of the smaller circle was called a *deferent*. The smaller circles around points on the deferent were known as *epicycles*. As the center of the epicycle moved along the deferent, the planet moved in a looping path. The concept of a loop explained the retrograde motion of the planets. (See Figures 1.7 and 1.8.)

The centers of the epicycles of Mercury and Venus were pictured as always being on a line connecting the earth and the sun. This explained the fact that these planets are seen only near the sun.

The Ptolemaic system is worthy of great respect. It was based on observations, and it was mathematically sophisticated. A theory or hypothesis need not be correct to be useful. The Ptolemaic system provided a framework for accumulating and organizing observational data. For centuries the data seemed to corroborate the hypothesis; it was possible to use the system to predict planetary motions and positions with acceptable accuracy. As observational data increased in quantity and quality, however, it became harder to fit the system to the data. The Castillian king Alfonso the Wise (1221–1284) remarked that if he had been present at the creation, he would have suggested a much

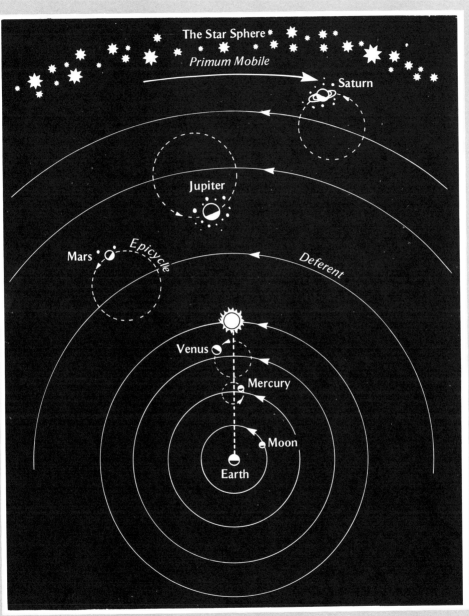

The Ptolemaic system. The centers of the epicycles of
Mercury and Venus lie on a line connecting the centers of
the earth and the sun.

FIGURE
1.7

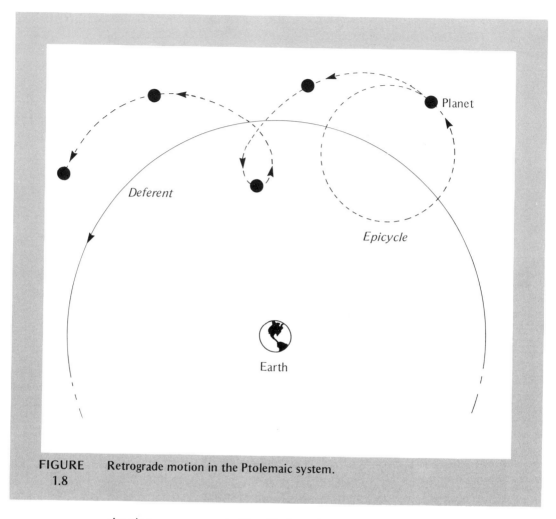

FIGURE
1.8

Retrograde motion in the Ptolemaic system.

simpler arrangement. The Ptolemaic system had such "common sense" appeal that it had come, unfortunately, to have a special sanction within Christian theology. It was felt that the Creator would certainly have placed man at the center of the universe.

The Copernican System

The Copernican system conceives of a spinning earth revolving about the sun. Although the concept was not entirely original with Nicolaus Copernicus, his work on the system makes its name appropriate.

A heliocentric (sun-centered) system had been proposed by several early thinkers, including the Greek Aristarchus in the third century B.C. It has already been mentioned that Copernicus had seen references to such thinking in the writings of Plutarch. However, these early proposals for a heliocentric concept were not accepted, and the Ptolemaic concept of a geocentric system dominated thought for many centuries.

Copernicus was a man of remarkably diversified training and accomplishments. He was born in what is now Torun, Poland, on February 19, 1473. He was the nephew of Bishop Lucas Watzelrode, who financed his education and later helped him get situated. He studied mathematics and astronomy at the University of Cracow. Later he spent eight years in Italy studying law, theology, medicine, the classics, and more mathematics and astronomy. Through the influence of his uncle, he became canon of Ermland and later canon of Frauenburg Cathedral. Official responsibilities kept him busy for several years with religious, business, and political affairs. He apparently met his heavy responsibilities effectively. He also became well known as a physician. (See Figure 1.9.)

FIGURE
1.9

Nicolaus Copernicus. (Yerkes Observatory photograph.)

Copernicus developed his concept of a heliocentric universe over a period of many years. Although his friends knew of his work, he did not publish it. Such a concept would have been heretical and especially inappropriate coming from a church official. A young mathematician named Georg Joachim Rheticus came to study with Copernicus in 1539. He remained with Copernicus, then an old man, for more than two years and helped persuade him to publish. Copernicus permitted Rheticus to take the manuscript to Nuremburg for printing. Copernicus was on his deathbed when the first printed copy was placed in his hands. It is said that he was semiconscious and could barely hold this book which was to have such profound influence on humanity's view of the universe.

Rheticus had to leave Nuremberg before the printing of the manuscript was completed, so he arranged for Andreas Osiander, a Lutheran minister, to oversee the remaining work. Osiander decided to add a preface which, he apparently thought, would make the book more acceptable to theologians. He also changed the title that Copernicus had given it, *Six Books on the Revolutions (De Revolutionibus, Libra VI)*, to *On the Revolutions of the Celestial Orbs*. Presumably this was to imply that the book considered only the sky and did not question concepts about the earth. Since Copernicus was unable to read the book after publication, many years passed before it was known that Osiander had made these confusing alterations.

Copernicus addressed his introduction of *De Revolutionibus* to Pope Paul III. He explained in the introduction that ancient writers (Plutarch, for instance) had caused him to think that the earth might move and that such eminent churchmen as the Cardinal of Capua and the Bishop of Kulm had encouraged him. Actually, his system did not come under intense attack from his own Roman Catholic church for several years; perhaps this was because *De Revolutionibus* was too mathematical to be easily read and understood. *De Revolutionibus* was not placed in the *Index of Forbidden Books* until 1616; but it was not removed until 1835. The earliest attacks came from Protestants. Martin Luther, a contemporary of Copernicus, said of him, "The fool wants to overturn the whole art of astronomy."

Although his contribution was great, Copernicus did not advance human understanding from the Ptolemaic system to where it is today in one giant step. Like Ptolemy, he conceived of many complex epicyclic motions. He hypothesized that the planets revolved about the sun (still in epicycles moving on deferents); that the moon revolved around the earth; that the retrograde motion of Mars, Jupiter, and Saturn resulted from the revolutions of the earth; that the daily rising and setting of celestial objects resulted from the earth's rotation; and that the outermost sphere of the fixed stars was motionless. (See Figure 1.10.)

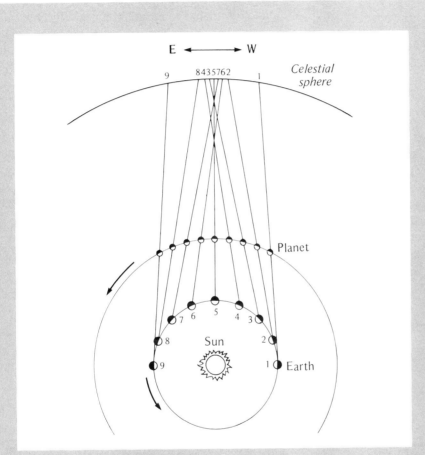

FIGURE
1.10

Retrograde motion in the Copernican system. As the earth moves from point 1 to 9, the planet is projected against the celestial sphere at points 1 through 9. The planet appears to move eastward until it reaches point 4. It appears to move westward on the celestial sphere from point 4 to 6 and then resume its eastward motion. (From D. Scott Birney, *Modern Astronomy*, 2nd ed., p. 20. Copyright 1974 by Allyn and Bacon, Inc.)

Copernicus's system was not actually simpler than Ptolemy's, although it has often been said to be, for it also required epicyclic motions. Nor did it permit a more accurate prediction of planetary motions; sometimes a planet would be as much as two degrees from the position predicted by either system. In addition to removing man from the center of the universe where he certainly appeared to be—and where the theologians believed God to have placed him—the system defied "common sense." Why didn't the spinning earth fly apart? Why weren't we swept away by a great wind? Why didn't the distant stars show a parallactic shift as the earth revolved around the sun? Copernicus answered that it is harder to picture the spinning celestial sphere remaining intact than to picture a smaller earth spinning more slowly remaining intact. He postulated that the air near the earth was carried along with the moving earth. He suggested that the stars were so distant that the relatively small diameter of the earth's orbit did not permit a parallactic displacement to be observed. (Parallactic shifting is the apparent change in position of an object when it is viewed from different places. A simple example of parallactic shifting is observed if you hold your finger a foot or so in front of your face and view it and the background first with one eye and then with the other.)

Despite the imperfections in his system, it is difficult to overstate the importance of Copernicus's contributions to intellectual and scientific history. He stimulated thought and work that resulted in gigantic achievements. He created a new concept of the universe and our place in it.

REFINING THE COPERNICAN SYSTEM

Although there were 14 centuries between Ptolemy and Copernicus, there were only 144 more years before refinements of the Copernican system brought an understanding of the main features of solar system design and mechanics. When Copernicus died in 1543, four important developments were still needed to achieve this understanding:

1. Additional and more accurate observational data on planetary positions had to be accumulated.
2. More accurate descriptions of solar system orbits, based upon observational data, were needed.

3. Some properties of falling and moving objects needed to be described and understood.

4. A theory was needed to account for the force controlling the motions of orbiting bodies.

Among many who contributed to the search for understanding were four celebrated giants: Tycho Brahe, Johannes Kepler, Galileo Galilei, and Sir Isaac Newton. Tycho gathered observational data. Kepler used Tycho's data to describe solar system orbits more exactly. Galileo made the first telescopic observations, popularized the heliocentric concept, and analyzed the motion of falling objects. Newton explained the motion of falling objects and orbiting planets using laws of motion and gravitational force.

Tycho Brahe (1546–1601)

Tycho Brahe (ordinarily referred to by his first name only) was one of the greatest astronomical observers of all time. Working in pretelescopic times, he designed large, accurate instruments for measuring the positions of celestial objects, and he used his instruments with immense patience and skill. Tycho was born to a noble Danish family, raised by a prosperous uncle, and educated at his uncle's expense at German universities. While he was a student in Germany he was involved in a bizarre incident. During a party at the home of a professor, he argued heatedly with another young Dane about a mathematical problem. A week or so later, while inebriated, they argued again and attempted to solve the problem with swords. For the rest of his life Tycho wore a silver nose painted to resemble skin. (See Figure 1.11.)

After he returned to Denmark from his studies in Germany, Tycho's interest and ability in science came to the attention of King Frederick II. From Elsinore Castle, Frederick had looked out at the island of Hveen and thought it would be a good place to establish an observatory. He offered Tycho the island and financial support for the rest of his life. It was a remarkably generous arrangement, and Tycho occupied the island in the grand manner. He built two observatories, Uraniborg ("Castle of the Heavens") and Stjerneborg ("Star Castle"). They included luxurious living quarters as well as the finest observatories of the day. He staffed them with many technical assistants.

FIGURE
1.11

Tycho Brahe. (Yerkes Observatory photograph.)

Tycho was unable to accept the Copernican system completely. He devised his own "compromise" system in which the sun and the moon revolved around the earth while the other planets revolved around the sun. The theological arguments against a sun-centered system influenced Tycho, but his chief reason for rejecting the concept of an orbiting earth was his inability to detect a parallactic shifting of the stars. He was confident that an observer on a moving earth would see the stars seem to shift as they were viewed from different places in the earth's orbit. His instruments—the finest available—were capable of measuring angles to an accuracy of two minutes of arc, and yet he could not detect an apparent shifting of the stars. He calculated that the fixed stars would have to be at least 100 times more distant from the sun than the outer planets if parallax were to go undetected, and this seemed unreasonable. Copernicus had said that this was indeed the case—the stars were too distant to seem to shift as the earth orbited the sun. (See Figure 1.12.)

FIGURE
1.12

A quadrant used by Johann Hevelius (1611–1687).
Hevelius was able to measure angles to an accuracy of one
minute of arc, surpassing the accuracy of Brahe's measurements. (Yerkes Observatory photograph.)

It is difficult to overstate Tycho's zeal for accurate work. For example, suspicious that Copernicus had been inaccurate to the extent of three or four minutes of arc in measuring the altitude of the sun at the winter solstice, Tycho sent an assistant to Fraunberg to remeasure it. Kepler had immense respect for Tycho's accuracy. After Tycho's death, Kepler worked for more than a year trying to determine Mars's orbit from Tycho's observational data. He finally decided that either his approach was wrong or Tycho's data was inaccurate; there was a discrepancy of eight minutes of arc in his calculations. He had such confidence in Tycho that he could not conceive of his being inaccurate to that extent. By accepting and perceptively analyzing Tycho's data he was able to make a great contribution to our knowledge of planetary motion.

After the death of Frederick II, Tycho had a less comfortable relationship with the Danish government. In 1599 he moved to Prague to work under the protection of Emperor Rudolf II. Young Kepler came to work under Tycho and was with him in the days before he died in 1601. Tycho left his observational data to Kepler.

Johannes Kepler (1571–1630)

Johannes Kepler was born in Wurttemberg. His first collegiate studies were for the Lutheran ministry. Although he became a giant in the history of science, he retained a keen interest in religion and was something of a mystic. His activities took him on strange tangents at times: he augmented his income by casting horoscopes; he defended his mother against charges of witchcraft; he promised his dying wife that he would marry one of her friends, who then refused him; he wrote an important mathematical paper after being puzzled by the volume of a barrel of wine bought at a harvest fair; he wrote *Somnium*, an early example of science fiction; he thought that the motions of the planets were related in some mystical way to musical harmonics. (See Figure 1.13.)

Kepler's patient and insightful analysis of Tycho's data led him to the discovery of the following three laws of planetary motion:

1. The planets move in elliptical orbits with the sun at one focus.
2. The line from the sun to a moving planet sweeps over equal areas in equal intervals of time.
3. The squares of the periods of revolution of the planets are proportional to the cubes of their mean distances from the sun.

FIGURE
1.13

Johannes Kepler. (Yerkes Observatory photograph.)

This constituted immense progress toward knowledge of the design of the universe and toward modern physical science. Carefully gathered data had been analyzed with mathematical precision, and the puzzling result (ellipses instead of circles) determined. The concept of a helio-centric solar system was still heretical to some, so Kepler's ellipses were not universally accepted. And it was not until 78 years later that Sir Isaac Newton would explain what force moved the planets in elliptical paths.

Galileo Galilei (1564–1642)

Galileo Galilei was Italian. Planning to become a physician, he was studying medicine at the University of Pisa in 1581. While attending mass in the cathedral there, he observed the swinging of a lamp hanging from the ceiling. He was surprised to note that the time (period) required for a complete swing appeared to be independent of the width of the swing (amplitude). He experimented and confirmed that the period of a pendulum's swing depends upon the length of the pendulum and not upon the amplitude of its swing. Apparently it was the observation of the swinging lamp that turned Galileo from medicine to the physical sciences; he became a founder of the field of mechanics, the study of forces and motions. He was a faculty member at the Universities of Pisa and Padua, and later he was a mathematician and philosopher for Grand Duke Cosimo II de' Medici. (See Figure 1.14.)

FIGURE
1.14

Galileo Galilei. (Yerkes Observatory photograph.)

The telescope was invented in Galileo's day, and he was the first to record observations of the sky. In 1609 and 1610 he made several highly significant discoveries with a thirty-power telescope he constructed. He observed that the moon is not a perfectly smooth sphere but rather has a rough and mountainous surface; he even estimated with reasonable accuracy the height of some lunar mountains from the length of their shadows. He discovered that the diffuse light of the Milky Way comes from a multitude of individual stars. He projected an image of the sun on a screen and saw that it was marked by spots; by observing the motion of the spots he concluded that the sun makes a complete rotation in about 27 days. He discovered four satellites orbiting about Jupiter. (We now know of fourteen.) He observed that Venus has a cycle of phases similar to that of the moon. This contradicted the Ptolemaic system. According to that system, the sun could never be between the earth and Venus, and Venus could never appear to be in the full or near full phase. (See Figures 1.15, 1.16, and 1.17.)

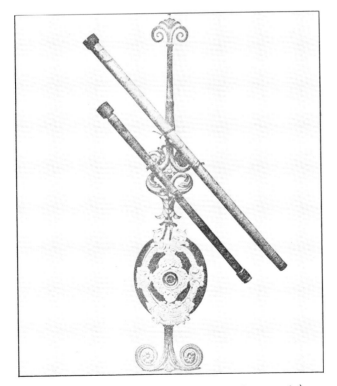

FIGURE
1.15

Galileo's telescopes. (Yerkes Observatory photograph.)

FIGURE
1.16

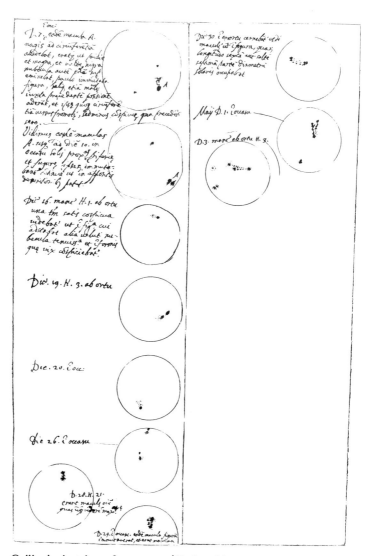

Galileo's sketches of sunspots. (Yerkes Observatory photograph.)

These observations made Galileo a firm advocate of the Copernican system. Kepler had been corresponding with him and trying to persuade him to accept this view. Beginning with his *Siderius Nuncius* (*The Starry Messenger*) in 1610, Galileo became a great popularizer of the Copernican system. He wrote in Italian rather than Latin, and his works were read widely. The Church objected to his views, and in 1616 he was

FIGURE
1.17

Galileo's sketches of Jupiter's satellites. (Yerkes Observatory photograph.)

ordered to stop teaching the Copernican system. He remained silent for only a few years. His classic work, *Dialogue Concerning the Two Chief World Systems*, was published in 1632. Soon after, the Inquisition ordered him to Rome, where he was forced to recant. However, he had made his impact on the mind of the people, and he had advanced our understanding of the universe.

Sir Isaac Newton (1642–1727)

Sir Isaac Newton placed the capstone on the structure that Copernicus had begun to construct. He was born in England in 1642, the year of Galileo's death. He entered Trinity College, Cambridge in 1661 as a student of mathematics. To escape the Black Plague, Newton went to the family farm in 1665. At the age of 24, and within a period of a few months, he developed the binomial theorem, differential calculus, the first two of his laws of motion, and important new concepts about light and optics; he also began to develop the formula for gravitational attraction. His creative work is one of the most awesome intellectual achievements in the history of mankind! (See Figure 1.18.)

FIGURE 1.18

Sir Isaac Newton. (Yerkes Observatory photograph.)

He returned to Cambridge, where he became a professor of mathematics and published important papers in optics. Perhaps because of a controversy with Robert Hooke, who claimed that he, not Newton, deserved credit for the inverse square law, he did not publish again for several years. Finally, prevailed upon by his friend Edmond Halley, he published *Principia Mathematica* in 1687. This is certainly one of the greatest of all scientific works. The book develops the theory of universal gravitation and demonstrates that the terrestrial laws governing Galileo's swinging lamp and the celestial laws governing Kepler's orbiting planets are all explained by the more encompassing law of universal gravitation.

Newton's greatness rapidly became evident. Many honors were awarded him, including knighthood and the presidency of the Royal Society. He died in 1727 at the age of 84.

Newton is reputed to have remarked modestly that if he had seen farther than other men it was because he had stood on the shoulders of giants. However, it is noteworthy that the epitaph on his tomb in Westminster Abbey states, "Mortals, congratulate yourselves that so great a man has lived for the honor of the human race."

MEASUREMENT

Measuring the Earth

Scientists have been curious not only about the design of the universe, but also about its size. The problem of determining the earth's dimensions is a difficult one, and measurements are still being refined. Eratosthenes, a Greek astronomer, made a remarkably successful attempt to measure the earth's circumference in about 200 B.C. He knew that on the first day of summer the sun was directly overhead at Syene, Egypt; it had been noted that the sun was directly above a vertical well in Syene at noon of that day. At the same time, he found that at Alexandria, 500 miles to the north, a vertical pole cast a shadow indicating that the sun was about 7.2 degrees from overhead. He assumed that the sun was far enough away that its light would come to the earth in essentially parallel rays. Since 7.2 degrees is $\frac{1}{50}$ of the 360 degrees in a circle, he reasoned that the circumference of the earth was 50 × 500 miles, or 25,000 miles. This result agrees well with modern measurements. (See Figure 1.19.)

Measuring the Universe

Determining the sizes and distances of celestial objects is obviously a massive problem. We will be extending and refining these measurements for years to come.

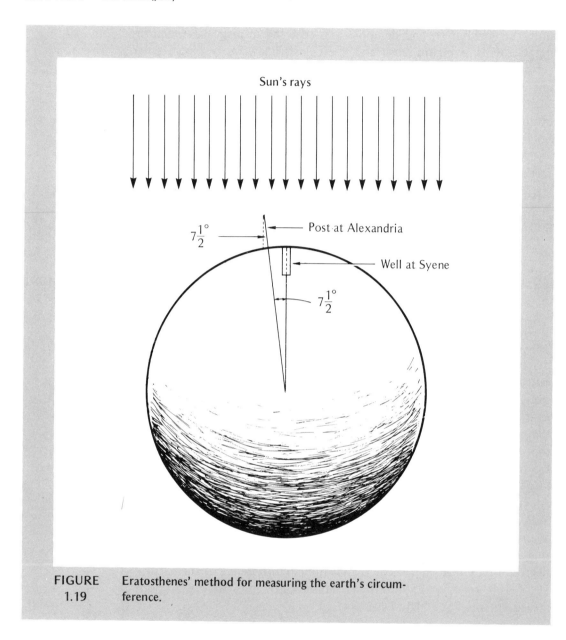

FIGURE 1.19   Eratosthenes' method for measuring the earth's circumference.

As long ago as 250 B.C., Aristarchus, a Greek astronomer, attempted to determine the relative distances to the sun and the moon from the proportions of the right triangle formed by the sun, the earth, and the first-quarter moon. His geometry was sound, but he had no means of measuring the angles with appropriate accuracy. He estimated that the sun was about 19 times as distant as the moon; it is actually about 400 times as distant.

As already indicated, one of Kepler's laws of planetary motions reveals a regular relationship between the periods of revolution of the planets and their distances from the sun. Since observation had already revealed the periods of the planets, Kepler was able to determine the *relative* distances of Mercury, Venus, Earth, Mars, Jupiter, and Saturn from the sun. If he had been able to determine the *actual* distance of any one of them from the sun, he would have known the actual distances of all of them. It was not until about 40 years after his death that a reasonably accurate determination of the distance of the earth from the sun was made.

The distance to the moon, which averages about 383,000 kilometers, has been known with reasonable accuracy for several hundred years. It can be determined through triangulation, using simple geometric principles. A surveyor can measure the distance to an inaccessible object by measuring a base line and sighting the object from each end of the base line. Knowing the angles formed with the base by the two lines of sight and knowing the length of the base, the surveyor can determine the remaining sides of the triangle formed by the lines of sight and the base line. The same technique can be applied to determine the moon's distance. Two different astronomical observatories mark the ends of the base line, and the moon is sighted from each at the same time. The necessary angular measurements are determined from the parallactic shift of the moon; as viewed from the two observatories, the moon will seem to shift in front of the background of more distant stars. (See Figure 1.20.)

It is much more difficult to apply this technique to determining planetary distances, because the parallactic displacements are so much smaller. In 1671 a French astronomer led a scientific expedition to French Guiana and sighted Mars from there. At approximately the same time Mars was observed from Paris. Measurements using this long base line revealed a slight parallactic shift, and it became possible to make a reasonable determination of the distance to Mars. Since Kepler had already determined the relative distances of the six known planets from the sun, determination of the distance of Mars from the earth made it simple to calculate the distances of all of the other planets from the sun. The distance of the earth from the sun was calculated to be 87,000,000 miles. Although this is short of the generally accepted 93,000,000 miles (150,000,000 kilometers), it put solar system dimensions in perspective.

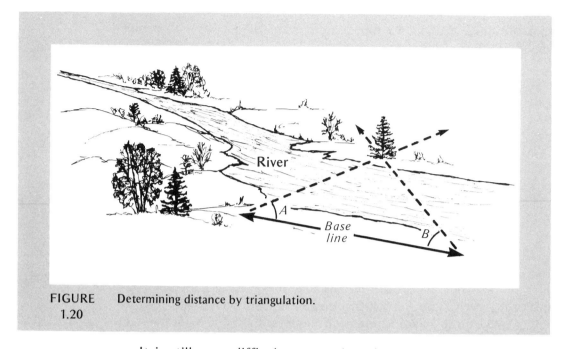

FIGURE
1.20   Determining distance by triangulation.

It is still more difficult to use triangulation methods to determine stellar distances. There is no earthly base line long enough to allow us to observe stellar parallax. The diameter of the earth's orbit itself must be used as a base line. When some of the closer stars are photographed at dates six months apart from opposite ends of a base line 300,000,000 kilometers long, they reveal very tiny parallactic shifts. The angles formed by the parallactic shifts are smaller than that subtended by the diameter of a dime viewed from a mile away!

The diameter of the moon and other objects close enough to appear as disks in telescopes can be calculated from their angular diameters, provided their distances are known. The geometry is simple; it is, in effect, the reverse of determining distance through parallactic displacement. As illustrated in Figure 1.21, one is determining the base line (diameter), knowing the apex angle and altitude of the triangle.

Additional methods for measuring the universe will be discussed later.

Locations on the Sky

It is necessary that we be able to describe positions in the sky precisely. Many aspects of astronomy and related fields are dependent upon this. Navigators, for instance, cannot determine their positions from the stars unless they know the positions of the stars exactly.

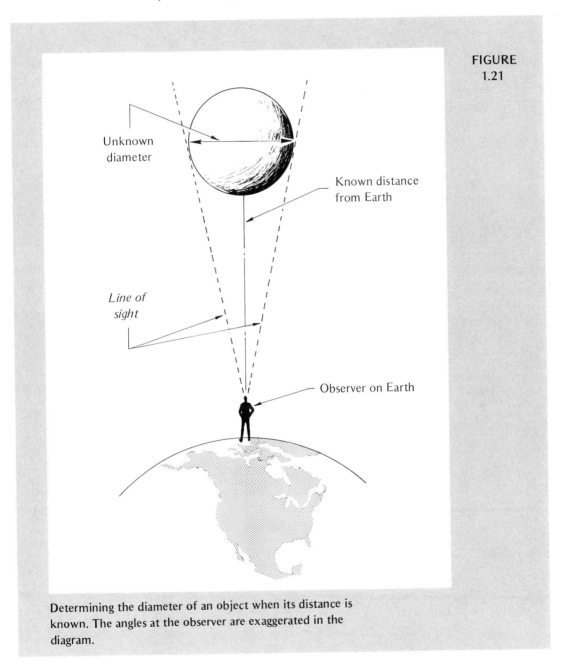

FIGURE
1.21

Unknown diameter

Known distance from Earth

Line of sight

Observer on Earth

Determining the diameter of an object when its distance is known. The angles at the observer are exaggerated in the diagram.

Positions on the celestial sphere can be identified in a way very similar to the one used to identify the latitude and longitude of points on the earth's surface. A coordinate system of imaginary lines is imposed on the sky. The *celestial equator* divides the sky into two hemispheres. If the equatorial plane of the earth were extended, it would intersect the celestial sphere at the celestial equator. If one were standing on the earth's equator, the celestial equator would be directly overhead; it would connect the eastern and western points on the horizon. The north and south *celestial poles* are directly above the north and south poles of the earth. Lesser circles are drawn parallel to and on each side of the celestial equator. These are *parallels of declination*, comparable to parallels of latitude on the earth's surface. Great circles are drawn perpendicular to the celestial equator passing through the celestial poles. These *hour circles* are comparable to meridians of longitude on the earth's surface. The coordinates of a star in this system are *declination* and *right ascension*. Declination, comparable to latitude, is measured north and south of the celestial equator in degrees and minutes of arc. Right ascension is the distance from the hour circle that passes through the *vernal equinox* to the hour circle that passes through the star. It is measured eastward in hours and minutes of time; one hour is equivalent to fifteen degrees of arc. The vernal equinox is the point at which the sun crosses the celestial equator at the beginning of spring. Right ascension is comparable to longitude. The declination and right ascension of Sirius, for instance, are 16 degrees 19 minutes South and 6 hours 43 minutes ($-16°19'$ and R.A. 6^h43^m). The *equatorial coordinate system* is illustrated in Figure 1.22.

Seasons

How centrally important in human affairs are the seasonal changes that the earth experiences! People had found ways of identifying the time at which seasons changed before they were able to identify the reason for the change. The giant stones in Stonehenge were placed on a British moor about 1650 B.C. in such a way that certain stones would be in alignment with the rising and setting sun on days of seasonal changes.

The sun's apparent eastward drift, resulting from the earth's revolution about the sun, has already been mentioned. The path along which the sun appears to move is the *ecliptic*. The axis of the earth is not perfectly upright in relation to the plane of its orbit; the axis is inclined

FIGURE
1.22

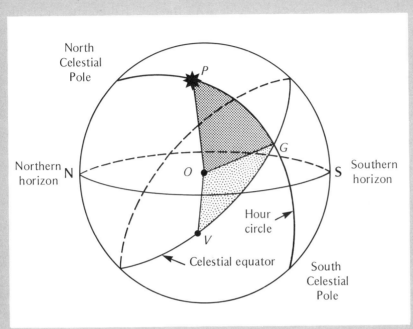

The equatorial coordinate system. The right ascension of
star *P* is *VG*—the distance from the vernal equinox eastward
along the celestial equator to the hour circle that passes
through the star. The star's declination is *PG*, its distance
from the celestial equator measured on the hour circle.

FIGURE
1.23

Stonehenge. (British Tourist Authority photograph.)

$23\frac{1}{2}$ degrees from perpendicular. As a consequence of this inclination, the ecliptic does not coincide with the celestial equator; it is inclined $23\frac{1}{2}$ degrees from the celestial equator. The ecliptic and the celestial equator intersect at two points. The points of intersection are the *vernal equinox* and the *autumnal equinox*. The points at which the ecliptic is farthest north and south of the celestial equator are the *summer solstice* and *winter solstice*. Summer and winter begin when the sun reaches the solstices—about June 22 and December 22. The sun is above the horizon for the longest and shortest times on the first days of summer and winter, respectively. Not only are there more hours of daylight in the summer, but the sun's rays strike the earth's surface more directly. Spring and fall begin when the sun reaches the vernal and autumnal equinoxes—about March 21 and September 23, respectively.

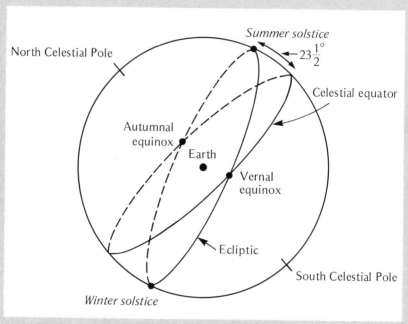

FIGURE 1.24

The changing of the seasons. The ecliptic and celestial equator intersect at the vernal and autumnal equinoxes; the solstices are 23½ degrees from the equator. Spring and fall begin when the sun is at the equinoxes; summer and winter begin when it is at the solstices.

On the days when the sun is at the equinoxes, it is above and below the horizon equal lengths of time, and it rises and sets due east and due west. The dates of the seasons in the Southern Hemisphere are the reverse of those in the Northern Hemisphere; for example, the Southern Hemisphere is most inclined toward the sun about December 22, and summer begins at that time. (See Figures 1.24 and 1.25.)

1963789

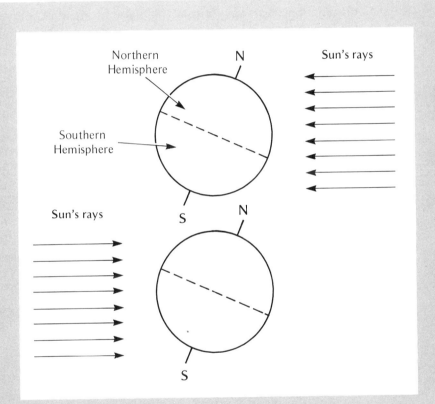

FIGURE
1.25

The concentration of the sun's rays. The axis of the earth is inclined 23½ degrees from the perpendicular toward the plane of the earth's orbit. When the sun is at the summer solstice, more of the sun's energy strikes each unit of surface area in the Northern Hemisphere than in the Southern Hemisphere, and the Northern Hemisphere experiences summer.

The Zodiac

The *zodiac* is a band of the sky through which the ecliptic runs. The zodiac is about 16 degrees wide; 8 degrees are on each side of the ecliptic. The moon, the sun, and the visible planets are essentially always seen within the zodiac, because of the flat pattern in which the moon and the visible planets orbit the sun. The zodiac is the region of the sky that holds special interest for those who believe in astrology. The zodiac is divided into twelve equal parts, beginning at the vernal equinox. These divisions are the *signs of the zodiac*, and they are named for the constellations that were within the divisions about 2,000 years ago. The names of the signs moving eastward from the vernal equinox are Aries, Taurus, Gemini, Cancer, Leo, Virgo, Libra, Scorpius, Sagittarius, Capricorn, Aquarius, and Pisces. The vernal equinox and the twelve signs have shifted noticeably westward in the past 2,000 years, and the signs no longer contain the constellations for which they are named. This was caused by the earth's *precession*, a kind of wobbling motion of the earth's axis. The earth's axis wobbles much like a spinning top. One of the results of this motion is that the equinoxes slide westward along the stationary ecliptic, completing the circuit in about 26,000 years. This effect is the *precession of the equinoxes*. Astrologers believe that the position of the sun, moon, and planets in the zodiac, particularly at the moment of one's birth, have a profound relationship to developments in one's life. For centuries thousands of people in all walks of life have taken astrologers seriously. Scientists generally attempt to discredit them.

Timekeeping

Not only must we accept the passage of time, but we must measure its passage accurately. Civilized people everywhere have used the smoothly turning sky as a great natural clock, and they have contrived mechanical clocks and other devices to run in harmony with it.

The interval in which the sky appears to turn once can be noted by observing two successive *transits* (passages) of a star or fixed point in the sky across the upper half of the *meridian*. The meridian is the great circle that passes through the *zenith* (the point on the celestial sphere directly overhead), the north and south points of the horizon, and the

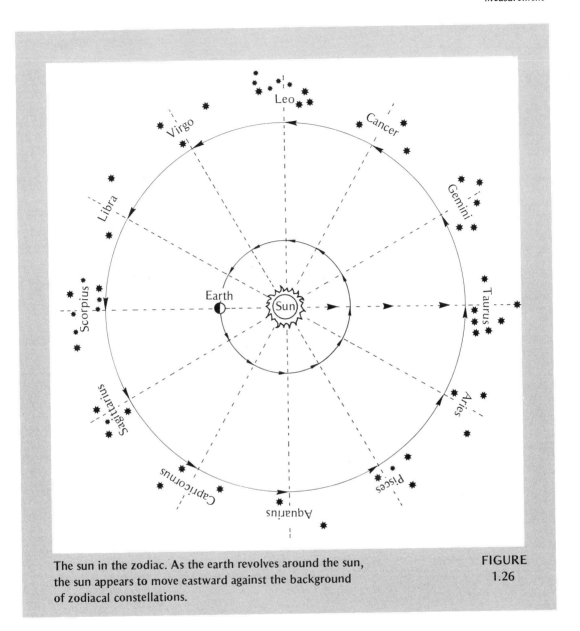

The sun in the zodiac. As the earth revolves around the sun, the sun appears to move eastward against the background of zodiacal constellations.

FIGURE
1.26

nadir (the point on the celestial sphere opposite the zenith). The interval between these transits is the *sidereal day*. Astronomers actually use the moment of transit of the vernal equinox, rather than a star, to designate the beginning of each sidereal day. Clocks keeping sidereal time are useful in observatories, because astronomers use sidereal time in locating objects. Astronomers find the right ascension and declination of a star in tables. They then need the sidereal time to tell them where the vernal equinox is in relation to the meridian, for right ascension is measured eastward from the vernal equinox. So, knowing sidereal time and right ascension, astronomers know whether or not the star is above the horizon (assuming that the declination of the star is such that it rises and sets at the latitude of the observer). (See Figure 1.27.)

FIGURE 1.27

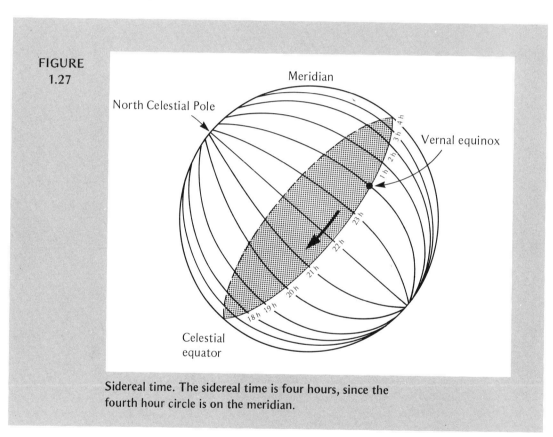

Sidereal time. The sidereal time is four hours, since the fourth hour circle is on the meridian.

Most of us do not find star time useful; we need a clock set so that the sun, not the vernal equinox, is near the meridian at noon. We need to keep solar time rather than sidereal time, since our daily activities are influenced so greatly by the sun. If we were to set an accurate clock at noon each day when the sun was actually on the meridian, we would find that the clock seemed to gain and to lose rather irregularly during the course of a year. This is because the earth moves around the sun in an elliptical orbit. It travels more swiftly when it is nearer the sun; thus, the sun too moves eastward along the ecliptic at a varying rate. The effect of this variation is eliminated through the use of *mean solar time*, which is based on the concept of a mean sun moving eastward along the celestial equator at a uniform rate. That rate is the average rate at which the real sun moves around the ecliptic in a year. The difference between mean solar time and apparent solar time at any instant is the *equation of time*. Sundials indicate apparent solar time. The shadow cast by the sundial's gnomon falls on the noon marker when the sun is on the meridian.

If all of us were to set our clocks to noon when the mean sun was on the meridian, this too would be unsatisfactory. Clocks a short distance to the east of us would be ahead, and those a short distance to the west would be behind. Clocks would differ by four minutes in places separated by one degree of longitude, since the earth rotates one degree in four minutes. This problem is solved by the use of *standard time*. Standard time zones are about 15 degrees of longitude wide, and all clocks within a particular zone are set to indicate the same time; at most places in the zone, the mean sun will be only approximately on the meridian at noon.

When we travel substantial distances to the east or west, we enter different time zones and must adjust our watches to keep local time. Adjusting our watch may not be enough; we may need to adjust our calendar. Travelers moving westward from New York to San Francisco must set their watches back an hour three times, because they enter three different time zones. Suppose a plane could leave New York and fly westward around the world at the speed of the earth's rotation at that latitude (some jet planes can fly faster than this). If the plane left New York at 9:00 a.m., it would pass over Chicago at 9:00 a.m., over Denver at 9 a.m., over San Francisco still 9:00 a.m., etc. At this rate, it would still be 9:00 a.m. when the plane returned to New York, but one day later than when it left. Where should passengers change their calendars to the next day? Travelers change dates when crossing the *International Date Line*. The International Date Line follows the 180th meridian of longitude through the Pacific Ocean, swerving in a few places to avoid islands and land masses. The date becomes one day later when the line is crossed traveling westward. If it is 9:00 a.m. on

Tuesday when one reaches the International Date Line, it is 9:00 a.m. on Wednesday after one crosses it. Calendars are turned back one day when the line is crossed traveling eastward.

Calendars

Our affairs require large units of time too, so we need units to measure longer intervals than days, hours, minutes, and seconds. Just as it has seemed natural to use the apparent turning of the sky to measure short intervals of time, the changing appearance of the moon and the passage of the seasons seemed to be the key to measuring longer intervals. Some difficult problems have resulted from the fact that the periods in which the sky appears to turn (day), the moon to go through its phases (month), and the four seasons to complete their cycle (year) are not exact multiples of each other. Through the centuries many attempts have been made to develop a satisfactory calendar.

The calendar year is the *tropical year*, the interval between two successive arrivals of the sun at the vernal equinox. The tropical year is 365 days, 5 hours, 48 minutes, and 46 seconds in length. This is about 20 minutes shorter than the *sidereal year* because of the earth's precession. The sidereal year is the interval in which the sun appears to make a revolution around the sky against the background stars.

In 45 B.C. Julius Caesar established a calendar that was quite similar to the one we use today. He knew that the year of the seasons (the tropical year) was about $365\frac{1}{4}$ days. He made this the average length of the calendar year by having each successive three years of 365 days be followed by a leap year of 366 days. Since the tropical year is about 11 minutes shorter than $365\frac{1}{4}$ days, the *Julian calendar* eventually became several days out of phase with the changing of the seasons. Pope Gregory XIII instituted the *Gregorian calendar*, a refinement on the Julian calendar, in 1582. He eliminated ten days from the calendar for 1582 (October 5–14), and he ruled that century years should be leap years only if evenly divisible by 400. The year 1900 was not a leap year, but the year 2000 will be. The Gregorian calendar was not adopted in England and the American colonies until 1752. Eleven days rather than ten (September 3–13) were dropped from the calendar in 1752. An average year in our present Gregorian calendar is about 26 seconds longer than the tropical year. Perhaps in the very distant future another adjustment will be necessary.

Celestial Navigation

Accurate timekeeping and a coordinate system that permits us to locate stars precisely on the celestial sphere made navigation by the use of the stars possible. If a line connecting a star to the center of the earth is visualized, the line will touch the earth's surface at a point directly beneath the star, the *substellar point or geographical position* of the star. When the navigator knows the geographical positions of two stars, they can serve as lighthouses do in coastal navigation. To know the geographical positions, the navigator must know the coordinates of the stars and the time, since the geographical positions move constantly as the earth turns beneath the stars. The right ascensions and declinations of conspicuous stars are recorded in tables for the navigator, who is also equipped with a very accurate clock, a chronometer. By measuring the altitudes of the two stars with a sextant and by noting the times at which the observations were made and the right ascension and declination of the stars, the navigator can determine his or her position.

CONCLUSION

The history in this chapter is a highly important facet of our scientific and intellectual heritage. It is more than astronomical history. These developments, cornerstones of modern science, have had profound influence on our view of ourselves and our relationship to the physical universe.

This brief history also provides some insight into the way scientific inquiry advances the frontiers of knowledge. The importance of careful observation as a basis of inquiry and the importance of freedom of inquiry should be evident. Also, these events should reveal the danger in trying to make physical phenomena matters of religious doctrine.

FOR REVIEW, DISCUSSION, OR FURTHER STUDY

1. Is there (as the title of one of the suggested readings might imply) a scientific basis to astrology?

2. What influence has astrology had upon art, literature, and music?

3. Who was Giordano Bruno? How and why did he die?

4. Why was the trial of Galileo important? Was Galileo "framed"?

5. Can the influence of the Copernican revolution be discerned in the opening sentence of the Declaration of Independence?

6. Suppose that the earth were to stop spinning on its axis but that its other motions were to continue. Which of the following statements would then be true?
   (a) We would have about six months of daylight followed by about six months of darkness.
   (b) The sun would rise in the west each morning and set in the east each night.
   (c) It would always be nighttime in the United States, but it would always be daytime on the other side of the earth.
   (d) It would always be daytime in the United States, but it would always be nighttime on the other side of the earth.

7. Suppose that the earth stopped moving around the sun, but that nothing else about its motions changed. (This is impossible, of course.) Which of the following statements would be true?
   (a) The sun would never rise or set.
   (b) The sky would not seem to turn in any way.
   (c) The sun would rise in the north and set in the south.
   (d) We would see the same stars on every night of the year.

8. At 8:00 p.m. tonight, we see the bright star shown below. Where will the star be at 8:00 p.m. one month from now? Will it be at 1, 2, 3, or 4?

9. We are looking at a bright star in the east. It is 10:00 p.m. Suppose we look at the star one month from now, but at 8:00 p.m. Where will the star be? Will it be at 1, 2, 3, or in the same place it is now?

10. Suppose we can see the moon in the south at sunrise. Where would the moon be in the diagram below—at 1, 2, 3, or 4?

SUGGESTED ACTIVITIES

If possible, it is best to make several simple nontelescopic observations of the sky before beginning to study Chapter 1. The total time needed for these observations is only a few hours spread over a month or so. The first chapter will be much more meaningful if you have made these observations with care. If it is not possible for you to make the observations before studying the chapter, you should begin them when you begin study of the chapter. Some of the observations must be made from the same location on two or more different nights; in these instances the positions of celestial objects are to be compared at various times in relation to fixed landmarks.

Observing the Turning Celestial Sphere

(a) First locate the North Star, Polaris. The Big Dipper, easily found in the northern sky, will aid you. Draw an imaginary line between the two stars in the bowl of the dipper opposite the handle. Extend the line about five times the distance between these two pointer stars, and it will lead you to the vicinity of Polaris. Then note the position of Polaris in relation to some landmarks (houses, trees, etc.). Observe the North Star one and two hours later and compare its position in the sky to its position at the time of the first observation. Compare the positions of Polaris at various times on several succeeding nights.

(b) Sketch the Big Dipper and the North Star in relation to landmarks. Sketch the positions of these objects one and two hours later on the same night. Label and save the sketches. Based upon the sketches, what can you conclude about the apparent motion of the Big Dipper during one hour, two hours, and twenty-four hours? Observe the positions of the Big Dipper twenty-four hours after making the first sketch to verify your conclusion.

(c) At the same hour of the night one month after your first observation of the Big Dipper, make another sketch of the Big Dipper and Polaris in relation to the landmarks. Compare the sketches drawn at the same hour of the night a month apart with those drawn two hours apart on the same night.

(d) On any evening, note the positions of a few conspicuous stars in the east, south, and west. Notice how their positions have changed one and two hours later. As one faces south, do the stars appear to move clockwise or counter-clockwise? As the hours pass, do the stars change their positions in relation to one another? Does a group of stars retain its shape as time passes?

(e) Consider the sky toward the south. Is there a "South Star" or an appropriate spot in the sky? If there is such a star or spot, how would you describe its location?

(f) Select a conspicuous star that appears fairly low in the west as soon after sunset as the stars become visible. Sketch its location in relation to landmarks. Note the position of the star soon after sunset on several nights during the following month.

(g) Follow the same procedure as in (f) with a star that appears low in the east just after dark.

Observing the Moon

Many calendars indicate the dates on which the moon will be in the new, first-quarter, full, and last-quarter phases.

(a) Keep track of the moon for a month by looking at it off and on during the day and night. Consider the location of the moon when it is in various phases. Is the waxing crescent that follows new moon seen in the east or west? In which phase is the moon when it is in the south in the early hours of daylight? Where is the first-quarter moon when darkness comes?

(b) Sketch the position of the moon in relation to some surrounding stars. Sketch its position in relation to these same stars on the next two or three evenings.

(c) Record the time at which the moon rises one evening. Compare this with the times at which it rises on the next two or three evenings.

(d) Note the cusps (points) of the waxing and waning crescent moons. Are they pointed toward or away from the sun?

Observing the "Moving" Sun

Watch for the appearance of stars low in the west as soon after sunset as possible. Visualize the position of the sun among the stars, even though the sun is below the horizon. Repeat this observation on a few evenings during the next month.

Consider the apparent motion of the sun against the background of stars. Does it appear to move eastward or westward against the background of fixed stars?

Observing the Wandering Planets

There are five visible planets: Mercury, Venus, Mars, Jupiter, and Saturn. Rarely can all be seen on the same night. Find out from your instructor which planets are currently in the evening or morning sky and how to locate them. Observe Venus and/or Mars; at least one of them is usually visible. This may require some observations in the predawn sky. Sketch the position of Venus and/or Mars against the background of fixed stars one evening (or morning) and at weekly intervals during the following month. At the end of the month, consider how the planets have moved in relation to the fixed stars.

SUGGESTED READINGS

Armitage, Angus. *Copernicus: The Founder of Modern Astronomy*. New York: A. S. Barnes and Company, Inc., 1962. 236 pages.

Bok, Bart J., "A Critical Look at Astrology," *The Humanist*, Vol. 35, No. 5 (September/October 1975), 6.

Bronowski, Jacob. *The Ascent of Man*. Boston: Little, Brown and Company, 1973. 448 pages.

Particularly recommended are Chapter 6, "The Starry Messenger," and Chapter 7, "The Majestic Clockwork."

Cohen, I. Bernard, "Galileo," *Scientific American*, Vol. 181, No. 2 (August 1949), 40.

Cohen, I. Bernard, "Isaac Newton," *Scientific American*, Vol. 193, No. 6 (December 1955), 73.

Gibson, R. E., "Our Heritage from Galileo Galilei," *Science*, Vol. 145 (18 September 1964), 1271.

Gingerich, Owen, "Copernicus and Tycho," *Scientific American*, Vol. 229, No. 6 (December 1973), 86.

Gauguelin, Michael. *The Scientific Basis of Astrology*. Translated from the French by James Hughes. New York: Stein and Day, 1969. 255 pages.

Hoyle, Fred. *Astronomy*. Garden City, New York: Doubleday & Company, Inc., 1962. 320 pages.

This is a scholarly but very readable history of astronomy by an eminent contemporary astronomer. The illustrations alone are most informative; it is an unusually handsome volume with more than 400 illustrations, many of which are in color.

Jerome, Lawrence E., "Astrology: Magic or Science?" *The Humanist*, Vol. 35, No. 5 (September/October 1975), 10.

Koestler, Arthur. *The Watershed: A Biography of Johannes Kepler*. Garden City, New York: Anchor Books, Doubleday & Company, Inc., 1960. 280 pages.

This paperback is one of the volumes in the Science Studies Series organized by the Physical Science Study Committee.

Lerner, Lawrence S., and Edward A. Gosselin, "Giordano Bruno," *Scientific American*, Vol. 228, No. 4 (April 1973), 86.

Levitt, I. M., and Roy K. Marshall. *Star Maps for Beginners*. New York: Simon and Schuster, 1964. 47 pages.

The charts in this book have a unique design, and they are remarkably easy to use and helpful for the beginner. There is a chart for each month, along with descriptive information about the appearance of the sky and some discussion of the mythology of the constellations.

Ley, Willy. *Watchers of the Skies*. New York: The Viking Press, 1963. 529 pages.

This is an unusually readable history of astronomy from Babylonian times to the space age.

Martin, Martha Evans, and Donald M. Menzel. *The Friendly Stars*. New York: Dover Publications, 1964. 147 pages.

Donald Menzel, a distinguished director of Harvard College Observatory, has revised this charming little book written by Martha Evans Martin in 1907. The book makes it easy to identify many of the bright stars and conspicuous constellations. It is one of the most helpful and interesting guides for the new stargazer.

North, J. D., "The Astrolabe," *Scientific American*, Vol. 230, No. 1 (January 1974), 96.

"Objections to Astrology: A Statement by 186 Leading Scientists," *The Humanist*, Vol. 35, No. 5 (September/October 1975), 4.

"Press Comment on 'Objections to Astrology'," *The Humanist*, Vol. 35, No. 6 (November/December 1975), 21.

Ravetz, J. E., "The Origins of the Copernican Revolution," *Scientific American*, Vol. 215, No. 4 (October 1966), 88.

Santillana, George de. *The Crime of Galileo*. New York: Time Incorporated, 1955. 371 pages.

Santillana, George de, "Greek Astronomy," *Scientific American*, Vol. 180, No. 4 (April 1949), 44.

"The Astrologers Reply," *The Humanist*, Vol. 35, No. 6 (November/December 1975), 24.

This is in response to the statement by 186 scientists, which is cited above.

FIGURE
2.1
The planet Earth. (NASA photograph.)

THE SOLAR SYSTEM

2

Man finally—and grudgingly—accepted the fact that the earth is not a stationary body at the center of a turning universe as observation, common sense, and theological considerations had made it appear. Where then is the earth in the universe? What is its place in the grand design of the cosmos? Such questions can be approached by first considering the solar system. The members of the solar system are the earth's immediate neighbors in the vastness of space.

The solar system is composed of the sun and a large number of objects that orbit it. As the sun moves through space, these objects are carried along, held captive in its gravitational field. The objects wheeling about the sun include nine planets and their thirty-four known satellites, thousands of asteroids, many comets, and myriads of meteoroids.

THE SUN

The sun dominates the solar system and provides the energy that sustains all life on earth. Yet it is a rather ordinary star. There are larger and smaller stars, hotter and cooler stars, denser and less dense stars.

The diameter of the sun is approximately 1,392,000 kilometers (864,000 miles), and its average distance from the earth is about 149,500,000 kilometers (93,000,000 miles). Its surface temperature is about 6,000 degrees Kelvin (10,000 degrees Fahrenheit); the interior may be as hot as 14,000,000 degrees Kelvin (25,000,000 degrees Fahrenheit). Spectroscopic analyses indicate that the sun is composed largely of hydrogen and helium. (The spectroscope is an instrument for studying light emitted by luminous bodies or absorbed by gases. It is possible to identify materials by determining the wavelengths of the light they emit or absorb.) The sun emits energy—because atoms of hydrogen fuse to form atoms of helium in a process akin to that which occurs in the hydrogen bomb. The sun rotates. Since the sun is in a gaseous rather than a solid state, the period of rotation is not equal for all parts of the sun's surface. Observation indicates that the sun rotates in about 25 days near its equator and in about 33 days near its poles.

The Photosphere

Since the sun is relatively close to the earth, its surface features can be observed. Being gaseous, the sun's visible surface is not distinct like the surface of a solid. The region of the sun that forms the apparent surface is the *photosphere*. The photosphere emits sunlight. If we look through telescopes equipped to protect the eyes, we can see that the photosphere is not uniformly bright. The edge of the disk, the *limb*, appears darker and redder than the center; this effect is referred to as *limb darkening*. The reason for this phenomenon is that the observer viewing the limb is not looking directly into the source of the light. The photosphere is dotted with small, relatively brighter *granules*, which are approximately 1,500 kilometers across. An individual granule forms and disappears in a matter of minutes. They are thought to be caused by hotter gases welling up from within the sun and then cooling somewhat at the photosphere. Other bright areas, larger than granules, are called *faculae*.

Sunspots

Sunspots are a particularly intriguing feature of the photosphere. Large sunspots can be seen without a telescope when the sun is viewed through haze. Sunspots were noticed centuries ago, but they received little attention until Galileo and his contemporaries directed the first telescopes toward the sun. Since then sunspots have been studied extensively. A great deal is now known about them, but much remains to be learned.

Sunspots are areas hundreds or thousands of kilometers across; they appear very dark because they are about 800 degrees Kelvin cooler than surrounding areas. Actually they are emitting light, and their dark color is a matter of contrast. Sunspots are not uniformly dark. The central region, the *umbra*, is darkest; it is surrounded by a lighter region, the *penumbra*, which is somewhat hotter. (See Figure 2.2.)

FIGURE
2.2

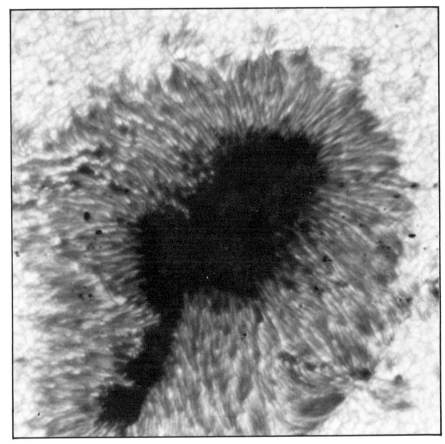

A sunspot's umbra and penumbra and the granular appearance of the photosphere in a telescopic photograph from a high-altitude balloon. (Project Stratoscope of Princeton University, sponsored by the Office of Naval Research, The National Science Foundation, and the National Aeronautics and Space Administration.)

One of the especially intriguing aspects of sunspots is the fact that their numbers increase and decrease cyclically. There is an interval of about 11 years between times when the number of spots is at a maximum. At times of maximum sunspot activity it may be possible to observe more than a hundred sunspots. During minimum activity there are hardly any spots.

Sunspots rarely occur singly. They usually develop as groups of small spots around a pair of large spots. An imaginary line between the pair of large spots would generally be parallel to the sun's equator. Some spot groups have lasted as long as three months and covered an area of several million square kilometers. (See Figure 2.3.)

FIGURE
2.3

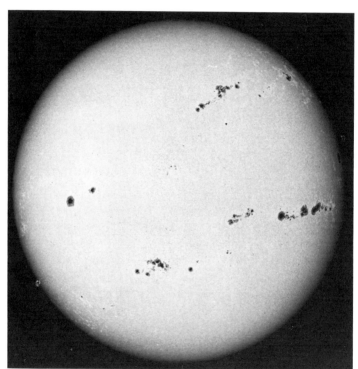

Photograph of the sun during a period of maximum sunspot activity. (Photograph courtesy of the Hale Observatories.)

Sunspots are not distributed randomly on the sun's surface. They usually occur within a band that stretches from 45 degrees north to 45 degrees south of the equator; the majority are located between 30 and 8 degrees north and between 30 and 8 degrees south. Sunspot activity develops in an interestingly symmetrical manner. At the beginning of a new cycle of activity, a few pairs of spots will develop at about 35

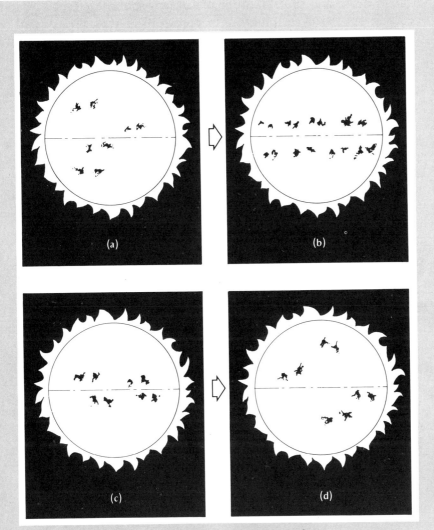

FIGURE
2.4

In (a), sunspot groups of a new cycle are beginning to form
at mid-latitudes while the last spots of the previous cycle
appear near the equator. In (b), about four years later than
(a), the spots have reached a maximum number and lie in
nearly parallel bands about 15 degrees north and south of
the equator. In (c), the spots, which are diminishing in
number, appear closer to the equator. In (d), eleven years
later than (a), the pattern of distribution is beginning to
repeat itself.

degrees north and south latitude. Then new spots develop closer to the equator. As the cycle progresses, more sunspots develop in two relatively narrow bands parallel to the equator and about equidistant from it. Near the end of the eleven-year cycle, only a few spot pairs are left, and these are in parallel zones just a few degrees north and south of the equator. At the same time a new cycle begins at latitudes of about 35 degrees north and south. (See Figure 2.4.)

A strong magnetic field is associated with each sunspot group. The polarity of these magnetic fields changes with the sunspot cycle. One of the large spots in each group is the positive pole of the magnetic field. The other spot is the negative pole. If the leading spot of a pair in the Northern Hemisphere (the spot toward the direction in which the sun is rotating) has positive polarity, then the leading spot in a Southern Hemisphere pair has negative polarity. In the next sunspot cycle the

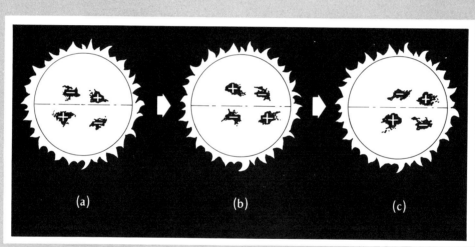

FIGURE 2.5   If, as in (a), the leading sunspot of a pair in the Northern Hemisphere has a positive magnetic polarity, the leading spot of a pair in the Southern Hemisphere will have a negative magnetic polarity. Eleven years later, in (b), this situation is reversed. In (c), the pattern of polarity is similar to that in (a), which was 22 years earlier. Thus a complete sunspot cycle is more properly regarded as a 22-year cycle than an 11-year cycle.

pattern of polarity is reversed. That is, leading spots in the Northern Hemisphere are negative, and leading spots in the Southern Hemisphere are positive. Thus, if the pattern of magnetism is considered, a complete cycle requires 22 years rather than 11. (See Figure 2.5.)

The cause of sunspots is not fully understood. It is believed that they result from the rotation and circulation of the sun's gases, which is influenced by the sun's magnetism. Some early astronomers thought that the sun had a dark surface surrounded by a shield of fire and that the spots were glimpses of the dark and cooler surface. Sir William Herschel (1738–1822), discoverer of the planet Uranus, even suggested that the surface might be inhabited.

There have been numerous attempts to correlate the eleven-year solar activity cycle with terrestrial phenomena. Many of the conclusions from these studies are of doubtful validity, but it does seem that the width of tree growth rings varies in a cycle that corresponds to the sunspot cycle. Perhaps a cyclical variation in solar energy influences earthly meteorological and climatic phenomena, which in turn influence life on earth.

The Chromosphere

The *chromosphere* is the layer of seething gases above the photosphere. It is about 12,000 kilometers thick. Two chromospheric phenomena are *prominences* and *flares*.

Solar prominences, which look like arching flames of gas, can be observed above the chromosphere. They are seen to best advantage along the limb of the sun. Formerly they became visible only when the moon passed between the earth and the sun and eclipsed the disk of the sun. Now there are instruments that make it possible to view prominences at other times. Prominences sometimes extend more than 300,000 kilometers above the chromosphere. Although some prominences do rise upward in a flamelike fashion from the chromosphere (*eruptive prominences*), more actually "condense" or develop above the chromosphere and flow downward into it. In some cases, the gaseous material of the prominence extends in a looping arch from one area of the chromosphere to another. The arched and looped prominences are believed to be shaped by the magnetic fields in which they occur. (See Figure 2.6.)

FIGURE
2.6

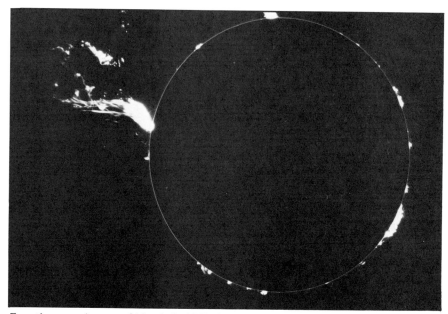

Eruptive prominence of March 1, 1969. It reached a height
of about 600,000 kilometers. (Photograph from the
Institute for Astronomy, University of Hawaii.)

Solar flares occur near sunspot groups. They seem to be violent
outbursts of hot gases, and they are much brighter than the surrounding
regions of the chromosphere. The term *flare* is appropriate, since these
eruptions reach their maximum brilliance in only a few minutes. Flares
disappear in an hour or so. (See Figure 2.7.) Prominences, on the
other hand, may last for many days. Both flares and prominences—
especially flares—can cause material to be propelled out into space and
leave the sun entirely. Flares hurl streams of ionized (electrically
charged) particles into space. Some of these particles reach the earth
a day or two after the flare. They cause the gases in our upper
atmosphere to glow, producing the *Aurora Borealis* ("Northern
Lights") over the Northern Hemisphere and the *Aurora Australis* over
the Southern Hemisphere. (See Figure 2.8.) These particles have an
adverse effect on the quality of radio reception. They diminish the
ability of our upper atmosphere (the ionosphere) to reflect radio waves
of some wavelengths. This reflection of radio waves is needed to permit
long-range reception over the earth's curved surface. It is possible that
the energetic particles from flares could be hazardous to astronauts

FIGURE
2.7

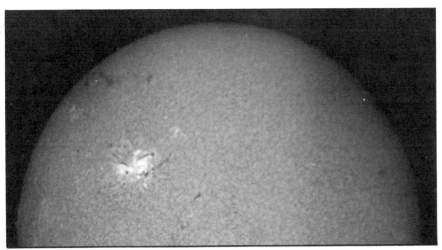

a

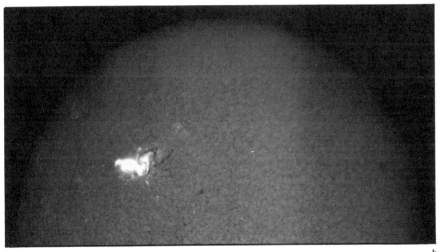

b

The solar flare of May 28, 1972. Photograph (b) was taken
40 minutes after photograph (a). They were taken with a
spectroheliograph, an instrument that utilizes only a carefully selected portion of the sun's light. (Photographs
courtesy of the McMath-Hulbert Observatory of The
University of Michigan.)

traveling in space. In order to anticipate the disruption of radio communications as well as the possible hazards to astronauts, solar activity
is monitored continually by a network of observatories.

FIGURE
2.8

Aurora Borealis photographed from an aircraft near Fort
Churchill, Canada. (NASA photograph.)

The Corona

The corona is the outermost part of the sun's atmosphere. It begins
where the chromosphere ends in an abrupt drop in the density of the
gases. The gas in the corona is much more rarefied than that in the
chromosphere. The speed of the particles constituting the very thin gas
of the corona is higher than that of the particles in the chromosphere,
which is to say that the temperature of the corona is higher than that of
the chromosphere. The temperature is nearly 2,000,000 degrees Kelvin
(about 4,000,000 degrees Fahrenheit). The particles are given their high
speeds (high temperatures) by explosive outbursts from the photosphere. At the speeds involved, the gas in the corona is ionized; that is,
it consists of electrically charged atomic particles. Since its gases are so
rarefied, little light is actually emitted by the corona. The light that
appears to be emitted by the corona is largely other sunlight scattered
by the minute particles of the corona. The shape of the corona changes
as the sunspot cycle progresses. It is rather round and flowerlike at
times of maximum sunspot activity; it is widened and flattened at the
poles during times of minimum sunspot activity. (See Figure 2.9.)

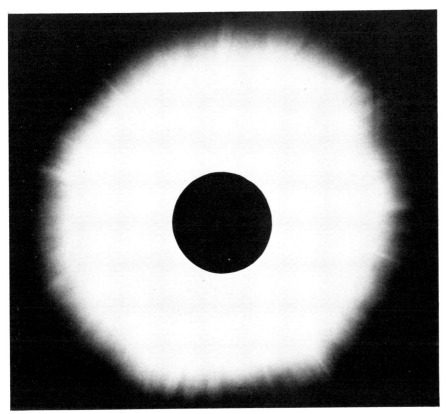

FIGURE
2.9A

The solar corona near a time of maximum sunspot activity.
The photograph was taken in Brazil during the eclipse of
May 20, 1947. (Yerkes Observatory photograph.)

The corona is ordinarily invisible because of the brilliance of the photosphere. It bursts into view at the moment of total eclipse, when the sun's disk is obscured by the moon. It is said to be an awesome sight. The *coronagraph*, a telescopic instrument, makes it possible to view the inner and brighter portion of the corona without an eclipse. It produces an artificial eclipse within the coronograph by obscuring the light from the sun's disk. Coronagraphs are located on mountain tops to minimize the disturbing scattering of light by atmospheric dust.

The visible corona extends for a million and a half kilometers or more outward from the sun's disk. Actually, since coronal particles are constantly streaming outward in all directions, the corona extends far beyond the portion that may be seen or photographed. In fact, this

FIGURE
2.9B

The solar corona near a time of minimum sunspot activity.
The photograph was taken in the Sudan during the
eclipse of February 25, 1952. (Yerkes Observatory
photograph.)

material extends beyond the earth, so the earth is, in a sense, within the
sun's corona. The stream of charged particles flowing outward from the
sun is known as the *solar wind*.

Solar Eclipses

A solar eclipse occurs when the moon passes directly between the sun and the earth. The moon then obscures part or all of the sun, and the moon's shadow touches the earth. A solar eclipse does not occur with every new moon, because the plane in which the moon orbits the earth is inclined slightly to the plane in which the earth orbits the sun. The two points at which the moon in its orbit passes through the plane of the earth's orbit are the *ascending node* and the *descending node*. The line connecting the two nodes is the *line of nodes*. Eclipses occur when the line of nodes approximately coincides with the line between the earth and the sun. During such an alignment, a solar eclipse will occur at new moon, and a *lunar eclipse* will occur when the full moon is darkened by the earth's shadow.

The disks of the sun and the moon appear to be of nearly equal size; that is, their *angular diameters* (angle at the observer's eye subtended by the diameter) are nearly equal (about 30 minutes of arc). However, the distance from the moon to the earth varies, and when the distance is more than average the moon's angular diameter is slightly less than that of the sun. At such times the moon causes an *annular eclipse*, for it is unable to eclipse the sun totally. In an annular eclipse a bright ring of the sun's disk surrounds the slightly smaller disk of the moon. (See Figure 2.10.)

The inner part of the moon's shadow, the *umbra*, receives no direct rays of the sun. It is a cone-shaped region of space which averages about 371,000 kilometers in length. In the remainder of the moon's shadow, the *penumbra*, only some of the sun's direct rays have been screened out. A total solar eclipse will occur only if the umbra touches the earth, and a total eclipse can be observed only from within the umbra. (See Figure 2.11.) In an annular eclipse, the tip of the umbra does not extend as far as the earth. A total eclipse can be observed only from a small strip on the earth's surface, for the spot formed by the umbra on the earth's surface is never more than about 265 kilometers in diameter and it is generally less. Because of the motions of the earth and the moon, the umbra sweeps over the earth's surface at a speed of several hundred kilometers per hour, tracing out a long, narrow *path of totality* from which the total solar eclipse can be viewed. Observers in a broad area on either side of the path of totality will be within the penumbra and will see a *partial eclipse*. In a partial eclipse the moon's disk obscures only a segment of the sun. The time from the beginning to the

FIGURE
2.10

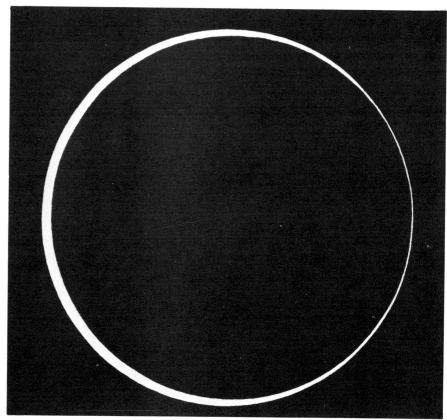

An annular eclipse of the sun, photographed from Senegal,
West Africa, on July 31, 1962. (Lockheed Solar Observatory, LMSC and U.S. Air Force photograph.)

end of an eclipse may be as long as about four hours, but the time of totality is about five minutes.

Total solar eclipses are not as rare as many believe; they occur about once every 18 months. However, since the path of totality is so narrow, opportunities to observe a total eclipse from a specified site are very rare. No total eclipses will be visible from the United States between February 26, 1979 and August 21, 2017. A total solar eclipse will not be seen from Europe until August 11, 1999.

The motions of the moon, earth, and sun are well understood now, and eclipses can be predicted with great accuracy. Since the motions of these bodies are regular, eclipses of various kinds are repeated in series.

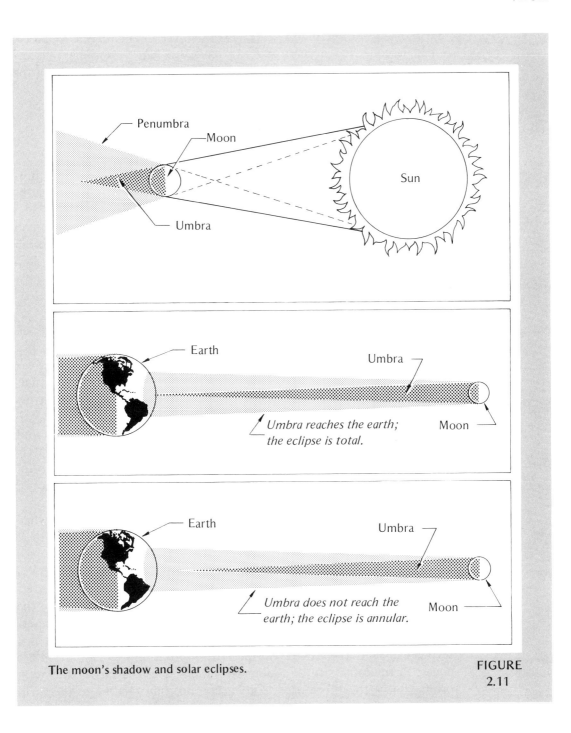

The moon's shadow and solar eclipses.

FIGURE
2.11

63

One of the intervals involved in reckoning series of eclipses is the *saros*, which is equal to 18 years and $11\frac{1}{3}$ days. Ancient astronomers discovered that similar eclipses recurred in these intervals and they used the saros to predict solar and lunar eclipses.

Invisible Radiation from the Sun

In addition to light, radiant heat, and solar wind particles, the sun emits radio waves, ultraviolet radiation, and x-rays. Our atmosphere absorbs and protects us from the ultraviolet rays and x-rays. Telescopes and other instruments must be lifted above the atmosphere by rockets in order to record the x-ray and ultraviolet radiation of the sun. *Radio telescopes*, radio receivers with special antennas, are used to study radio waves from the sun and other cosmic sources.

THE MOON

The moon is one of 34 known satellites revolving about the nine planets. It is an unusual satellite in that it is large compared to the planet about which it orbits; the diameter of the moon is about 3,476 kilometers (about 2,160 miles) and the diameter of the earth is about 12,727 kilometers (about 7,910 miles).

Despite the fact that the moon seemed hopelessly inaccessible prior to the 1960s, a great deal was known about it. Because the moon is relatively close to the earth, many details of its surface had been observed, photographed, and mapped. Many of its large craters had been named in honor of important scholars (Plato, Copernicus, etc.), and some of its mountain ranges had been given the names of terrestrial mountain ranges (Alps, Apennines, etc.).

The distance to the moon averages about 383,000 kilometers (239,000 miles). This ranges from about 356,319 kilometers when the moon is closest to the earth (at perigee) to about 406,594 kilometers when it is farthest away (at apogee). The moon rotates on its axis as it revolves around the earth. The periods of rotation and revolution are equal, so the same side of the moon always faces the earth. The moon orbits the earth and returns to the same position relative to the earth and the sun in about $29\frac{1}{2}$ days. The speed at which the moon revolves around the earth varies (it moves faster when it is closer to the earth), whereas the speed at which it rotates on its axis remains nearly constant. This enables us sometimes to see a bit farther around either the eastern or western edge of the moon. Since the axis of the moon is inclined slightly in relation to the plane of its orbit, sometimes the moon's north pole is tipped somewhat toward us and at other times the south pole is inclined toward us. These rocking effects are termed *librations*; they permit us to see a total of about 59% of the moon's surface at various times.

The Surface

Lunar landscapes are spectacularly harsh. The dusty, rocky surface seems white or gray against a black sky. There is no water or air. The lowlands are scarred by cliffs and fissures of various kinds. Mountain peaks tower to heights of more than 7,600 meters (25,000 feet). The ubiquitous feature is the numerous craters which indent the surface in countless places. (See Figure 2.12.)

FIGURE
2.12

Lunar landscape: a portion of the crater Copernicus.
(NASA photograph.)

Gravity on the moon's surface is only about one-sixth that on earth, and this is insufficient to hold an atmosphere. Without the moderating effects of an atmospheric blanket, temperatures on the moon's surface range from about that of boiling water to about 125 degrees Kelvin.

Craters

There are innumerable craters. They range in diameter from a few centimeters to 150 kilometers or more. Some appear to be much younger and more rugged than others. Substances—perhaps lava or dust—seem to have covered and smoothed the floors of some craters, such as Clavius, to form *walled plains*. Sometimes newer craters are seen on top of older ones.

There has been considerable controversy about how the moon acquired its pitted surface. Most astronomers originally believed that the craters were the result of the impact of meteoric bodies. Arizona's Meteor Crater seems to have been formed in this way. The chief opposing view was that the craters were volcanic in origin. Currently it is believed that craters were formed in both ways. Increasing knowledge of the moon's surface supports the belief that the moon has experienced volcanic activity, but most craters have the physical characteristics of bomb craters which are impact craters. (See Figure 2.13.)

FIGURE
2.13

Craters on the moon. (Photograph courtesy of the Hale Observatories.)

Some craters, such as Tycho, have patterns of long, narrow streaks radiating outward, sometimes for hundreds of kilometers. These are called *rays*; they are best seen at full moon, when they appear brighter than the surrounding surface. They follow the topography of the surface, they are thin, and they do not cast shadows. Rays are believed to have been formed from dust ejected from a crater at its formation and thrown for great distances in the low lunar gravity. (See Figure 2.14.)

FIGURE
2.14

Ray systems around the craters Copernicus and Kepler.
(Lick Observatory photograph.)

Maria

The first astronauts on the moon landed in Mare Tranquillitatis (the Sea of Tranquillity). Early observers thought that the dark areas (the face of the "man in the moon") were bodies of water and named them *maria*, meaning "seas" in Latin. Actually, the maria are plains and are relatively smoother than other areas. It appears that the smooth plains were formed by material—principally lava—that collected in lower regions and buried other surface features.

Rilles

The telescope revealed hundreds of crevasse-like features on the lunar surface many years ago. These are known as rilles. Typically, they are several kilometers long and less than a kilometer wide. Most of them are straight, but they are sometimes curved. Their variety has made it difficult to suggest their origin. Some seem clearly to be the result of faulting. The five unmanned Lunar Orbiter missions of the mid-60s transmitted 1,950 photos to earth which revealed a vast amount of surface detail too fine to be visible with a telescope. The *sinuous rilles* are a puzzling type of feature discovered by use of the orbiters. They resemble meandering river valleys on earth. Indeed, some have suggested that they were formed by flowing water. The Apollo 15 astronauts landed at Hadley Rille, a sinuous rille about 100 kilometers long, 1,500 meters wide, and 400 meters deep. Their observations suggest that the rille was formed by the collapse of a lava tube. (See Figure 2.15.)

Mascons

Spacecraft orbiting the moon have experienced an unexpected increase in velocity while passing over some of the maria. This suggested that concentrations of mass lie on or beneath the surface and exert gravitational attraction upon spacecraft passing overhead. These concentrations of mass are referred to as *mascons*; their nature has not yet been determined. Some have suggested that beneath the maria lie huge chunks of stone or metal which created the maria when they crashed into the moon. Others have suggested that water trapped in rocks

FIGURE
2.15

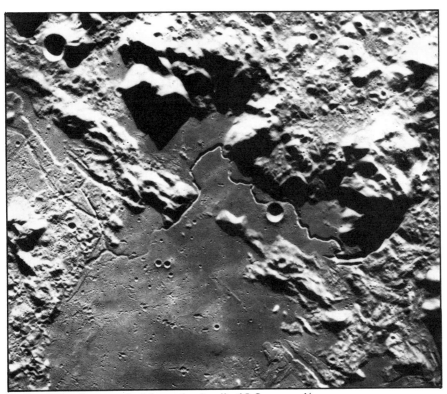

Hadley Rille photographed from the Apollo 15 Command/
Service module in lunar orbit. (NASA photograph.)

beneath the maria increases their density. Many believe that the mascons are formed of lava in which contraction has produced rock of
greater density than surrounding material. A material of greater density
would naturally have greater gravitational attraction for the spacecraft.

Composition and Formation

When the Apollo 11 astronauts, the first men to land on the moon,
returned to earth on July 24, 1969, they carried about 22 kilograms of
lunar material. Few scientific reports have been as eagerly awaited as
the analyses of these 48 pounds of moon rocks.

The samples included basaltic igneous rock, microbreccias (a mixture of soil and small rock fragments that have been compacted into

rock), and lunar soil. The samples showed pitting and local melting, which result from high velocity impact by tiny particles. The most common minerals found were pyroxene, plagioclase, ilmenite, olivine, and cristobalite. Three new minerals were found. The same elements that we know on earth were found, of course, but in different proportions. Sodium and potassium were less abundant in the basaltic rock samples than in crustal rocks on earth; titanium and zirconium were present in concentrations surprisingly higher than those here. The soil contained some elements, such as traces of gold, that were presumed to have been brought to the lunar surface in meteorites.

It was a futile hope that these first lunar samples would settle the controversy between "hot-moon" and "cold-moon" scientists. Cold-moon theorists hold that the moon was never—or at most was only briefly—hot and that it does not have a molten core. Some of them suggest that a cold moon might have been formed by the accretion of meteoritic material. Cold-moon advocates picture a lunar surface uninfluenced by lava flows. The oldest rock found on the earth is from lava flows of about 3.5 billion years ago. Cold-moon theorists hoped that the lunar samples would prove to be older than this—perhaps even as much as 4.6 billion years old. Meteoritic material striking the earth has contained rocks that old, and this has been held to be the probable age of the solar system.

Isotopic studies of the soil and microbreccia samples suggest that they might have come from rocks as much as 4.6 billion years old. However, basaltic crystalline rocks from the Sea of Tranquillity were found to be about 3.7 billion years old. They apparently crystallized from molten material nearly a billion years after the formation of the moon. A few months after the Apollo 11 mission, Apollo 12 astronauts returned to the earth with another 34 kilograms of lunar samples. A rock that they collected was found to be 4.6 billion years old. The determination was made by measuring the rock's content of rubidium-87 and strontium-87. The strontium isotope results from the radioactive decay of the rubidium isotope, which proceeds at a known rate; the relative abundance of the two isotopes indicates the period of time since the rock solidified. Other Apollo 12 samples seemed to have melted more recently than the 3.7 billion-year-old Apollo 11 samples. Cold-moon scientists are trying to adjust their theories to accommodate some melting, perhaps from heat generated by meteoritic impact, but most astronomers now find it hard to picture a moon that was never molten.

Scientist-astronaut Harrison H. Schmitt working beside a
huge boulder on the lunar surface during the Apollo 17
mission—the final Apollo mission—in December 1972.
(NASA photograph.)

FIGURE
2.16

Subsequent Apollo and Soviet missions have gathered more lunar
rock samples and observational data. It appears that there was a great
deal of melting on the moon near the time of its formation, perhaps 4.6
billion years ago, and that a second period of less widespread melting
ended about 3.1 billion years ago. The moon has apparently been
essentially inactive since. The more recent melting evidently caused lava
to flow into basins to form the maria. Seismic data suggest that the
moon has a layered crust about 60 kilometers thick; this is about twice
the average thickness of the earth's crust. Rock from the lunar high-
lands, about 70% of the moon's surface, is typically anorthositic gabbro
rich in plagioclase, a mineral containing aluminum and silicon. Younger

rock from the maria is basaltic and rich in iron. A third principal type of lunar rock is a radioactive basalt which has been found in limited areas, particularly Mare Imbrium and Oceanus Procellarium. Many lunar rocks display residual magnetism; the cause of this magnetism is undetermined. Geophysical considerations suggest that the magnetism was caused by the rotation of a molten core early in the moon's development. This is in conflict with the more common belief, derived from geochemical considerations, that heating of the moon began on the outside and moved inward—the heating was caused by the energy of the material that was accreting to form the moon. Some who believe the core was never molten say that the magnetism results from the moon having been in an external magnetic field; others say that the magnetization is the result of the shock from meteoritic impacts. Exploration of the moon is one of our greatest technological triumphs, but it has generated as many scientific questions as it has answered. We still cannot be confident about any explanation of when and how the moon was formed.

Phases

The phases of the moon are its several different apparent shapes as seen from the earth. These were described in Chapter 1. Sunlight always illuminates one-half of the moon, except in the unusual situation where the earth is directly between the sun and the moon, thus preventing sunlight from striking the moon. As the moon orbits the earth we see different portions of its lighted half, ranging from none of it at new moon to all of it at full moon. Figure 2.17 illustrates the relative positions of the sun, moon, and earth as the moon carries out its cycle of phases. The cycle of phases is completed in about $29\frac{1}{2}$ days. That is, the period between two new moons or two full moons is about $29\frac{1}{2}$ days.

Actually, the moon revolves about the earth in approximately $27\frac{1}{3}$ days rather than $29\frac{1}{2}$ days. This period of $27\frac{1}{3}$ days, the *sidereal month*, is the time required for the moon to orbit the earth and return to its original position against the background stars. However, while the moon is orbiting the earth, both the moon and the earth are also orbiting the sun. The $29\frac{1}{2}$ day period, the *synodic month*, is the time required for the moon to complete its cycle of phases and return to its original position relative to the sun and the earth.

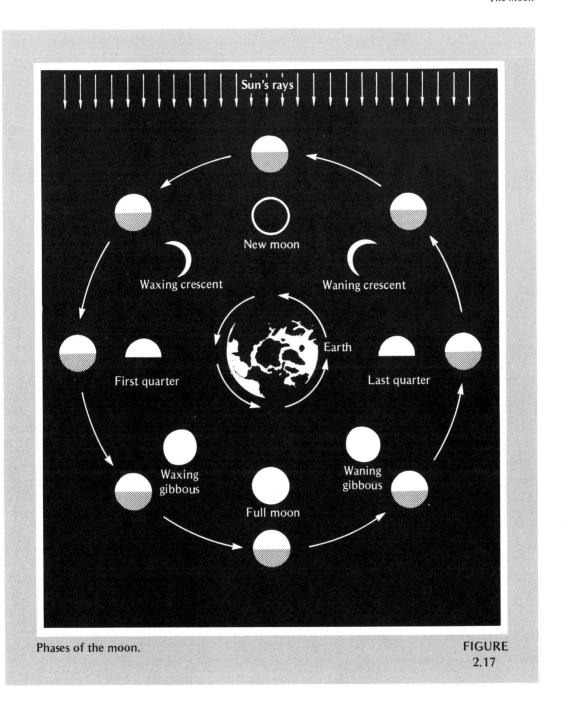

Phases of the moon.

FIGURE
2.17

73

During the crescent phase it is sometimes possible to see the entire disk of the moon faintly illuminated in addition to the small, bright crescent. This is caused by *earthlight*. At that time, the earth viewed from the moon would be in the full phase. The bright, full earth reflects enough sunlight to the moon to illuminate it slightly.

Much folklore involves the supposed correlation of the moon's phases with other phenomena. For instance, some almanacs advise readers to plant vegetables that bear crops above ground during the light of the moon (when it is waxing, that is) and vegetables that bear crops below ground during the dark of the moon (when it is waning).

Rising Times

The moon moves eastward against the background stars as it orbits the earth. Consequently, each day the moon rises later than the day before. Because the inclination of the moon's orbital path to the horizon varies, the daily delay or retardation in rising varies from about ten minutes to about an hour and a half. The daily retardation is least in the autumn, since that is when the inclination is least; that is, the moon's path is then most nearly parallel to the horizon. The full moon that occurs closest to the time when the sun is at the autumnal equinox is the *harvest moon*. On nights near the time of harvest moon, the lessened daily delay in the time of moonrise is very noticeable, and the nearly full moon can be seen rising only a little later on each of several successive evenings.

The Moon Illusion

The *moon illusion* causes the moon to appear larger when it is near the horizon. Since the angular diameter of the moon always remains about 30 minutes of arc, the moon illusion is clearly a perceptual phenomenon. Considerable scholarly work has failed to result in a universally accepted explanation of the moon illusion. Some believe that the perception of size is influenced by the position of the moon in relation to the position of the viewer's body. Others hold that the presence of other objects in the field of vision influences perception of the moon's size.

Tides

The rhythmic ebbing and flowing of tides has always been an important phenomenon in the lives of seamen and others who live and work on seacoasts. Changing depths and currents of water influence how and when these people can conduct many of their activities.

The gravitational attraction between the earth and the moon is chiefly responsible for tides; the attraction between the earth and the sun is also partially responsible. In considering the reasons for tides, it is helpful to imagine the earth as a sphere completely covered by a layer of water of uniform depth, as in Figure 2.18 (a). Since the gravitational attraction between bodies diminishes as the distance between them increases, the gravitational attraction between the moon and the water on the near side of the earth is greatest, causing a piling up or bulging of the water at N in Figure 2.18 (b). Because the solid earth is closer to the moon than the water on the far side of the earth, the earth is drawn away from the water there, causing bulging at F. Because some water moves toward N and F, the depth of the water is diminished at L and L¹. (These effects are all exaggerated in the figure.)

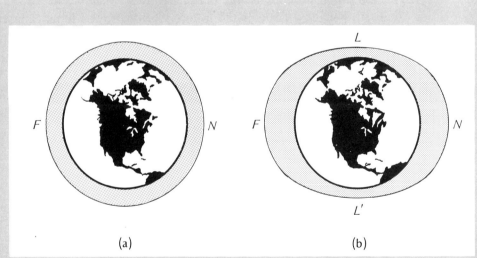

(a)　　　　　　　　　　　　　　(b)

Tides. The water in the oceans does not surround the earth to a uniform depth, as in (a). Rather, there are high tides on the sides nearer (N) and farther (F) from the moon. There are low tides at the points between: L and L′. The relative depths are greatly exaggerated.

FIGURE
2.18

The depth of the water at any point in an ocean changes as the earth rotates; the depth is greater when a given part of the ocean is at N or F than when it is at L or L'. The changing depth goes unnoticed in the open sea, but it is very apparent along the coasts. The depth of water may vary more than 40 feet between high and low water, depending upon the configuration of the shoreline.

A given seacoast on the rotating earth will be in the same relative alignment with the moon at intervals of about 24 hours and 50 minutes rather than 24 hours. For a given coast, then, there are two high tides and two low tides over the course of about 24 hours and 50 minutes.

When the moon, earth, and sun are in a relatively straight line (at new moon and full moon) the tidal effects are greatest; tides at these times are called *spring tides*. Tidal effects are least when the sun, moon, and earth form approximately a right angle (at first and last quarter); tides at these times are called *neap tides*.

Lunar Eclipses

The full moon is eclipsed when the earth is between it and the sun and the earth's shadow darkens the moon. Although total lunar eclipses do not occur any more frequently than total solar eclipses, when they do occur they are visible from the entire half of the earth facing the moon rather than from a narrow path of totality. The umbra of the earth is so large that there can be no annular eclipses of the moon, but partial lunar eclipses do occur.

The earth's shadow on the moon always has a curved edge. It was this evidence that convinced some early thinkers that the earth was spherical. They reasoned that only a spherical object could always cast a round shadow.

THE PLANETS

When people finally realized that the planets were nonluminous bodies orbiting the sun just as the earth does, they began to wonder whether the planets might be inhabited. Many were convinced that Mars in particular was inhabited. Several successful unmanned flights have begun planetary exploration. It appears that nearly all of the living things we

FIGURE
2.19

Views of the harbor at Cutler, Maine at high and low tide.
(Maine Department of Marine Resources photographs.)

know of could not survive on the planets, but hope of discovering life on Mars or elsewhere in the solar system has not been abandoned.

The known planets, listed in order of distance from the sun, are Mercury, Venus, Earth, Mars, Jupiter, Saturn, Uranus, Neptune, and Pluto. They may be categorized as either *terrestrial planets* or *giant planets*. The terrestrial planets are Mercury, Venus, Earth, Mars, and probably Pluto. They are so considered because in density, size, number of satellites, and period of rotation they are somewhat like the earth. The four giant planets (Jupiter, Saturn, Uranus, and Neptune) are much larger though less dense, they rotate more rapidly, and generally they have more satellites than the terrestrial planets. This is evident from the data summarized in Figure 2.20.

| | Mean distance from the sun | | Period of revolution | Equatorial diameter (kilometers) | Number of satellites | Period of rotation | Mean density (water = 1) | Mass (earth = 1) |
	Astronomical units	Millions of kilometers						
Mercury	0.387	57.9	88.0 days	4,867	0	58 days 16 h	5.46	0.055
Venus	0.723	108.1	224.7 days	12,109	0	243 days (retrograde)	5.23	0.815
Earth	1.000	149.5	365.26 days	12,754	1	23 h 56 m 04 s	5.52	1.000
Mars	1.524	227.7	687.0 days	6,787	2	24 h 37 m 23 s	3.93	0.107
Jupiter	·5.203	777.8	11.86 years	142,700	14	9 h 50 m 30 s	1.33	318.0
Saturn	9.539	1,426	29.46 years	121,000	10	10 h 14 m	0.69	95.2
Uranus	19.18	2,868	84.01 years	47,000	5	10 h 49 m	1.56	14.6
Neptune	30.06	4,495	164.8 years	51,000	2	16 h	1.54	17.3
Pluto	39.44	5,900	247.7 years	5,600(?)	0	6 days 9 h 17 m	5(?)	0.11

FIGURE 2.20   Planetary data.

Mercury

Mercury is difficult to observe. Since it never rises or sets more than about 90 minutes before or after the sun, it is generally obscured by the glare of the sun. The best telescopic views reveal no surface detail—only some indistinct shading of surface color. Because no distinct surface features can be seen, it has been difficult to determine the period of rotation. For many years it was thought that Mercury's periods of rotation and revolution were both 88 days. Study of radar reflections from the planet now indicates that Mercury rotates in 59 days. An unusual consequence of these periods of revolution and rotation is that one Mercury day is equal to two Mercury years. That is, Mercury will orbit the sun twice between two successive noons at a particular spot on the surface.

The planet has too little mass and gravitation to hold more than a trace atmosphere. The intense solar radiation probably produces surface temperatures of more than 600 degrees Kelvin; when darkness comes temperatures on the airless surface probably plummet to 150 degrees Kelvin.

The Mariner 10 unmanned spacecraft passed Mercury at close range three times between March 1974 and March 1975. This space mission provided numerous clear photographs of Mercury's surface and data about its weak magnetic field and trace atmosphere. The surface of Mercury appears to be much like that of the moon. However, one feature peculiar to Mercury is its numerous scarps or cliffs, often more than a kilometer high and 450 kilometers long. (See Figures 2.21 and 2.22.)

FIGURE
2.21

Mercury's surface photographed by Mariner 10 from a distance of 35,000 kilometers. The large flat-floored crater is about 80 kilometers in diameter. (NASA photograph.)

FIGURE
2.22

Mercury's surface photographed by Mariner 10 from a
distance of 6,000 kilometers. Craters as small as 150 meters
across can be seen. (NASA photograph.)

Venus

Venus can be brilliantly beautiful. Next to the sun and the moon it is
the brightest celestial object—bright enough to cast a shadow on a
moonless night and bright enough to be seen in broad daylight. It is the
conspicuous "evening star" sometimes visible in the west after sunset
and the "morning star" sometimes seen in the east before sunrise.

Venus is blanketed by a thick, heavy atmosphere which obscures
the surface from view and creates some horrendous conditions by
earthly standards. The atmosphere, composed largely (97%) of carbon
dioxide with small amounts of nitrogen (2%), water vapor, and ammonia, exerts a crushing pressure—about 90 times greater than atmospheric
pressure on earth. Beneath the dense atmosphere, the surface is only
dimly lighted. The atmosphere produces the "greenhouse effect"; it
admits radiation from the sun more readily than it permits the reradia-
tion of energy. Surface temperatures reach nearly 800 degrees Kelvin,
which is sufficient to melt some metals. Several unmanned U.S. and
Russian space flights to Venus have furnished information about conditions there. One of the Russian spacecraft transmitted data suggesting
that the surface at its landing point was covered by loose material with
a density of about 1.5 grams per cubic centimeter and a composition
somewhat similar to that of terrestrial granite. In October 1975 another

spacecraft survived the heat for 53 minutes and transmitted to earth the first television pictures of the Venusian surface. The surface in the vicinity of the spacecraft appeared to be covered with jagged rocks.

Since the Venusian surface is obscured from view, it was very difficult to determine the rotational period of Venus. Measurements with reflected radar rays indicate a rotational period of 243 days. The period required for Venus to orbit the sun is 224.7 days. Surprisingly, Venus was found to rotate in a clockwise direction. The typical direc-

FIGURE
2.23

Venus photographed by Mariner 10 from a distance of 720,000 kilometers: This is a mosaic of photographs taken with ultraviolet light. (NASA photograph.)

81

tion of revolution and rotation for solar system objects is counterclockwise (as viewed from above).

Important data about the structure of the turbulent atmosphere of Venus were provided by the Mariner 10 spacecraft in February 1974. (See Figure 2.23.)

Mars

In the past people were convinced they could see seasonal changes in vegetation on Mars along a network of canals constructed by intelligent beings. Excitement about Martian canals arose when lay people concluded more from an astronomer's reports than he intended they should. Giovanni Schiaperelli, an Italian astronomer, reported seeing straight lines on Mars in 1877; he used the Italian word *canali* (channels), which many misinterpreted as meaning manmade canals. Percival Lowell, a wealthy Bostonian, was stimulated by Schiaperelli's reports to establish an observatory in Arizona to search for life on Mars. The canali remained controversial, however, because many observers were unable to see them.

An exciting advance in our knowledge of the Martian surface came from the unmanned Mariner fly-bys. Three such space probes in the period from 1964 to 1969 transmitted 222 photographs covering about 20% of the planet. They revealed a heavily cratered surface much like the moon's. Mariner 9 went into orbit around Mars in November 1971. During the next several months it transmitted 7,329 photographs, many revealing football field-size features, plus data related to atmospheric and surface temperature and composition. The unmanned Viking I landed in July 1976 and transmitted telecasts of the Martian landscape and inconclusive results of tests for signs of life processes in soil.

Mars—unlike the "dead" moon—is a dynamic planet with a surface that continues to change. The surface is spectacularly rugged and varied with differences in elevation as great as 15 kilometers. One giant chasm is as large as the Rift Valley in Africa. There seem to be no folded mountain chains, but Nix Olympica—one of several volcanic mountains—is 500 kilometers wide at its base, 8 kilometers high, and has a crater at the top 65 kilometers across. The circular basin Hellas is 1,000 kilometers in diameter—twice as large as the moon's Mare Imbrium. There are numerous channels, apparently formed by streaming fluid (lava?

water?); some are tiny and others are a kilometer wide and thousands of kilometers long. (See Figures 2.24 and 2.25.)

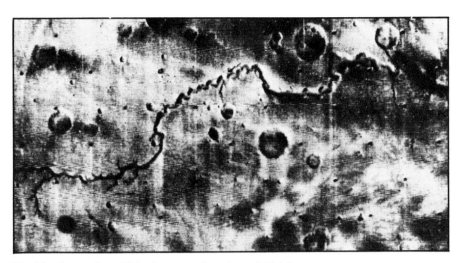

FIGURE
2.24

The Martian surface. This sinuous valley about 370 kilometers long and 5 kilometers wide was photographed from Mariner 9. (NASA photograph.)

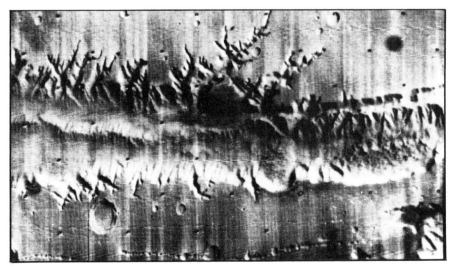

FIGURE
2.25

The Martian surface. This vast chasm with branching canyons eroding the adjacent plateaulands was photographed from Mariner 9. (NASA photograph.)

The inclination of Mars's axis is comparable to the earth's; therefore there are seasonal changes on Mars. Temperatures range from about 300 degrees Kelvin near the equator to about 175 degrees Kelvin in the polar regions. There are white polar caps which increase and decrease in size as winter and summer come to a particular hemisphere. These caps seem to be very thin layers of frozen carbon dioxide; they also seem to include water ice in much smaller amounts. The atmosphere is composed of carbon dioxide (95%), nitrogen (2.5%), argon (1.5%), and a little water vapor and oxygen; atmospheric pressure is vastly less than on earth. Blowing sand causes great dust storms. The color of some surface areas lightens and darkens alternately; this is believed to be the result of sand being blown onto and off of the surfaces. The Martian surface appears to have been altered greatly by volcanism, faulting, meteoritic impact, and erosion by thermal effects, wind, and fluids. It appears that there is no liquid water on Mars now, but the development of some of the surface features is most easily explained in terms of the former presence of running water. (See Figure 2.26.)

FIGURE
2.26

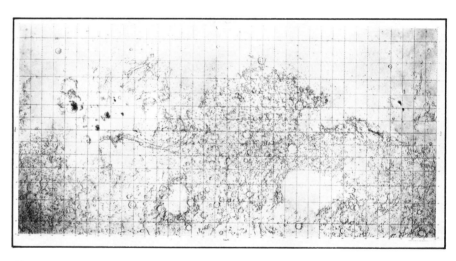

Shaded relief map of the Martian surface prepared from mosaics of Mariner 9 photographs. The map covers an area about 18 times that of the continental United States. (U.S. Geological Survey.)

Mars has two tiny satellites, Phobos and Deimos. These orbit Mars at distances of roughly 10,000 and 22,000 kilometers in periods of about $7\frac{1}{2}$ and 30 hours, respectively. It is an interesting coincidence that 150 years before their discovery, Jonathan Swift, writing in *Gulliver's Travels*, described two tiny Martian satellites orbiting close to the planet in periods of 10 hours and $21\frac{1}{2}$ hours. (See Figure 2.27.)

FIGURE 2.27

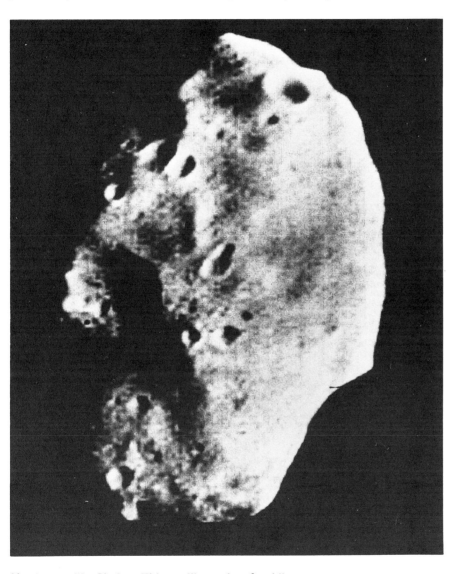

Martian satellite Phobos. This satellite, only a few kilometers in diameter, was photographed by Mariner 9. (NASA photograph.)

Pluto

Pluto is the planet farthest from the sun; it is perpetually dark and cold. The sun is nearly six billion kilometers from Pluto, so from Pluto it would appear to be the ordinary star that it is. Temperatures on Pluto are perhaps less than 150 degrees Kelvin. Its orbit is the most eccentric among the planets; at times Pluto is actually nearer to the sun than Neptune.

Because of its great distance and small size, Pluto appears as little more than a point of light in most telescopes. It has been difficult to determine its diameter and period of rotation. In the largest telescopes its image does subtend a discernible angle, and Pluto's diameter is believed to be less than 6,000 kilometers. Recent observations of periodically fluctuating light from Pluto suggest that it rotates in a little more than six days.

Percival Lowell and others became convinced that the gravitation of an unknown planet beyond Neptune was influencing the motion of Uranus. Clyde Tombaugh, a young astronomer at Lowell's observatory in Arizona, continued the search that Lowell had begun years before. He patiently compared pairs of photographic plates, looking for something with a changing position among the fixed stars. In 1930 he discovered Pluto. It seems appropriate that the first two letters of the name given to the planet are the initials of Percival Lowell. (See Figure 2.28.)

FIGURE
2.28

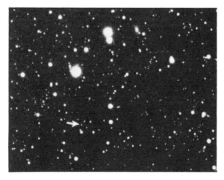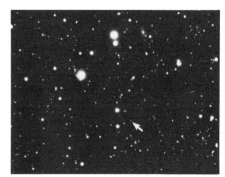

Portions of the photographic plates that led to the discovery of Pluto. The left photo was made on January 23, 1930; the right photo was made on January 29, 1930. The arrows show that a closer object (Pluto) has changed position against the background stars. (Lowell Observatory.)

Jupiter

Jupiter, the largest planet, differs greatly from the terrestrial planets. Its surface is perhaps an ocean of liquid hydrogen enclosing a core of solid material, which is also largely hydrogen. The surface cannot be seen through the thick atmosphere. The pressure at the surface, exerted by the crushing weight of the atmosphere, is more than one hundred thousand times as great as that on earth. Temperatures at the surface may be as high as 1,500 to 2,000 degrees Kelvin.

In view of these facts, it may seem incongruous to state that conditions on Jupiter may be conducive to the creation of life. Spectro-scopic studies indicate that the Jovian atmosphere of hydrogen also contains the simple chemical compounds methane and ammonia. Many believe that these materials were among key ingredients from which complex organic compounds were synthesized during the evolution of life on earth. Experiments in which mixtures of methane, ammonia, and water vapor are subjected to ultraviolet radiation or electrical discharges have produced some of the complex compounds present in living tissue.

Jupiter's atmosphere has a characteristic banded appearance visible even with a small telescope. Frequent changes in the appearance of the bands suggest great turbulence. A famous feature of Jupiter's atmo-sphere is the *Great Red Spot*. The Great Red Spot is oval in shape and covers several hundred thousand square kilometers. It has been ob-served for more than 200 years and has changed in appearance and location. It is thought that the Great Red Spot may be a very long-lasting hurricane. (See Figure 2.29.)

Jupiter enigmatically radiates twice as much energy as it receives from the sun. After a flight from earth of 21 months, the Pioneer 10 spacecraft passed within 130,000 kilometers of Jupiter in December 1973. It passed through radiation belts in which it was exposed to radiation equal to more than one thousand times the dosage ordinarily fatal to humans.

Jupiter is larger than all eight other planets combined. It rotates the fastest—once in about 10 hours. Because of this rapid rotation it is noticeably flattened at the poles. It has the most satellites (14), two of which are considerably larger than the moon.

It is interesting to observe the larger satellites transit the disk of the planet or be eclipsed by Jupiter's shadow. Observations of such eclipses permitted Ole Römer (1644–1710) to conclude that the speed of light

FIGURE
2.29

Jupiter's Red Spot, a shadow of the satellite Io, and Jupiter's cloud structure, photographed by Pioneer 10 from a
distance of 2,500,000 kilometers. (NASA photograph.)

was finite. He noted that the intervals between successive eclipses of a
particular satellite were not uniform. He noted that the interval was
several minutes longer when the earth was on the opposite side of the
sun from Jupiter (in opposition) than when the earth was between the
sun and Jupiter (in conjunction). He concluded that the additional time
was that needed for light to travel across the earth's orbit, and he
estimated that it took light 11 minutes to travel the earth–sun distance.
(The modern value is about eight minutes).

Saturn

Saturn, the ringed planet, presents a striking telescopic view. The rings
are not solid. They are composed of myriads of small particles. Spec-

troscopic studies suggest that the particles may be coated with frost. There are three concentric rings; the main ring has a less conspicuous ring on each side of it. The total diameter of the ring system is about 270,000 kilometers, but the rings are probably no more than 15 kilometers thick. They are so thin that stars can sometimes be seen through them. Viewed edge on, they are very difficult to see because of their thinness. (See Figure 2.30.)

FIGURE 2.30

Various aspects of Saturn's rings. (Lowell Observatory photograph.)

Saturn is similar to Jupiter. It is large and low in density, rotates rapidly, is accompanied by many satellites, and has a banded appearance. Titan, the largest of Saturn's ten satellites, has a diameter of 4,800 kilometers and possesses an atmosphere; no other satellite in the solar system is known to have an atmosphere.

Uranus and Neptune

Mercury, Venus, Mars, Jupiter, and Saturn are visible without a telescope and have been known since ancient times. Uranus is generally invisible, but sometimes appears to be one of the faintest visible stars. It was not discovered until 1781, when Sir William Herschel identified it as a planet.

Located far away from the sun, Uranus is extremely cold—perhaps 150 degrees Kelvin. Spectroscopic studies reveal hydrogen and methane in its atmosphere, but no ammonia. At the temperatures there, ammonia would solidify out of the atmosphere. Uranus is so far away that the telescope reveals relatively few details. It has a greenish color. The planet is apparently lying on its side, with its axis of rotation nearly in the plane of its orbit. Its period of rotation is believed to be a little less than 11 hours.

Neptune resembles Uranus in a number of respects. Its existence and location were predicted before the planet was discovered. It became evident that the orbital motion of Uranus was not entirely in accordance with theory. Some astronomers became convinced that Uranus was being influenced by an unknown and more remote planet. In 1846 Urbain Jean Leverrier, a French astronomer, calculated where the unknown planet should be. He asked another astronomer to look for it, and within an hour the new planet was found in the predicted location. More than a year earlier, John C. Adams, a young British mathematician, had made a similar calculation and prediction; unfortunately, he was unable to persuade an astronomer to look for the planet. Leverrier and Adams are now both credited with the discovery.

ASTEROIDS

There are many thousands of *asteroids* or *minor planets* revolving around the sun mainly between the orbits of Mars and Jupiter. Ceres, the first to be identified, was discovered in 1801. The nature of this body was puzzling for several months; Guiseppe Piazzi, its discoverer, thought at first that he had found a comet. Several astronomers thought he had found the "missing planet" which Johann Bode had predicted would be found between Mars and Jupiter. Bode had formulated a series of numbers that approximated rather closely the distances of the planets from the sun in astronomical units (an *astronomical unit* is the mean radius of the earth's orbit, which is approximately 150,000,000 kilometers or 93,000,000 miles). This numerical relationship, which is sometimes referred to as *Bode's law*, suggested that a "missing planet" should be at 2.8 astronomical units from the sun. Ceres revolves at 2.8 astronomical units from the sun.

Only a few more asteroids were located in the 50 years following the discovery of Ceres. After photographic techniques were developed, the discovery of asteroids became commonplace, because on photographic plates orbiting asteroids leave trails against the backgrounds of fixed stars. Asteroids are numbered consecutively as their orbits are determined; about 1,700 have been numbered. The discoverer of an asteroid has the privilege of naming it, so most asteroids have names in addition to their numbers. One can speculate about the inclusion of such names as Chicago, Scheherazade, Deborah, Elly, and Fanatica.

Ceres, the largest asteroid, has a diameter of about 800 kilometers. Pallas, Vesta, and Juno, with approximate diameters of 480, 400, and 200 kilometers, are the next largest. Since only a few minor planets display measurable disks in large telescopes, the sizes of most asteroids must be estimated from their brightness, assuming that all have equal reflectivity. Most asteroids are less than 40 kilometers at their greatest dimension. Asteroids are invisible to the unaided eye, except Vesta, which is occasionally visible.

Many of the asteroids fluctuate in brightness in periods of from about 1 to 18 hours. This fluctuating reflectivity of sunlight suggests that they are irregularly shaped and that they are rotating. Indeed, in a close approach to the earth (25,000,000 kilometers), the asteroid Eros was observed to be about 25 kilometers long and 8 kilometers thick and to be rotating in about $5\frac{1}{4}$ hours around an axis perpendicular to its long dimension.

The orbital periods of most asteroids are about $4\frac{1}{2}$ to 5 years. Although most of them orbit between Mars and Jupiter, some have markedly elliptical orbits; Hermes has passed within 1,000,000 kilometers of the earth. Some people speculate about the possibility of a collision of the earth and an asteroid. Although such an event is possible, the odds against it are literally astronomical.

Perhaps the asteroids themselves—or at least all but the few largest, which appear spherical—are fragments resulting from collisions. Since the total mass of the thousands of asteroids is only a small fraction of that of the moon, there is no indication that a major planet was broken up in their formation.

COMETS

Comets can be among the most awesome of natural phenomena, and most people see only a few in a lifetime. For centuries comets were widely regarded as evidence of the supernatural; they were viewed with fear and taken as a sign of famine, war, or pestilence. The ancient Chaldean astronomers had two divergent explanations for comets: one view held that they moved among the planets but were generally too far from the earth to be visible; the other considered them atmospheric phenomena caused by fires in violently rotating air. Unfortunately, the view that they were some kind of atmospheric phenomena prevailed, and only for the past few hundred years have they been correctly understood to be solar system objects.

Edmond Halley (1656–1742) was the first to apply Newton's work to comets in order to determine their orbits. He believed that the comet of 1682 was the same object that had appeared in 1607 and 1531; he predicted that it would appear again in about 76 years. We can again expect to see this comet, which now bears his name, in late 1985. History records its appearance on schedule as far back as 240 B.C.

Most comets seem not to be on such short period schedules; they seem to make a single appearance and then disappear forever. Actually, it may be that their orbits are so elongated that they become visible only at intervals of hundreds of years, and their schedules may not yet be evident. Halley's comet follows an elliptical orbit which takes it out beyond Neptune at times and to within one-half an astronomical unit of the sun at perihelion (*perihelion* is the closest point to the sun in a solar orbit). The orbits of some comets may be highly elongated ellipses which take them far out beyond Pluto. Another possibility is that those comets that seem to make only a single appearance in the vicinity of the sun may have parabolic or hyperbolic orbits, both of which are open-ended curves. These open-ended or seemingly open-ended orbits are often inclined sharply to the general plane of the solar system. Many comets do not revolve in the same direction as the planets.

Several comets revolve in periods of about six years and at aphelion are not much farther from the sun than Jupiter (*aphelion* is the farthest point from the sun in a solar orbit). They all revolve in the same direction as the planets, and their orbits lie nearly in the general plane of the solar system. They seem to have been partially "captured" by Jupiter's gravitation; thus these fifty or so comets are known as *Jupiter's family of comets.*

A few comets are discovered each year, often by amateurs who search patiently for them. Most of these are visible only with a telescope. Truly spectacular comets appear only a few times in a century.

A comet generally consists of a *coma,* a diffuse cloud, surrounding a faint *nucleus.* The coma, varying in size with distance from the sun, is often as big as Jupiter. The nucleus is only a few kilometers in diameter. In addition to the coma or head of the comet, a tail structure of extremely tenuous, less luminous gas and dust particles streams away from the head for millions of kilometers. The tail always points approximately away from the sun. The solar wind and radiation apparently blow the tail away from the head and the sun. (See Figure 2.31.)

FIGURE
2.31

Comet Tago-Sato-Kosaka, photographed from Cerro Tololo, Chile on January 11, 1970. The comet is at a distance of about 74,000,000 kilometers in this photograph; the length of the tail is about 10,000,000 kilometers. (Photograph courtesy of Freeman D. Miller and The University of Michigan.)

Comets have little mass, despite their size, and they are not hot, even though they are luminous. The mass of a comet is perhaps only 1/100,000 that of the earth, and most of this is in the tiny nucleus. Some of the light from a comet is reflected sunlight, and some is light emitted by the comet after absorbing energy from the sun.

The nucleus is thought to consist of frozen substances such as water, methane, ammonia, and carbon dioxide mixed with rocky and metallic material. One astronomer has likened the nucleus to a dirty snowball. As the comet approaches the sun, heat from the sun vaporizes some of the nucleus to form the coma and the tail. The rocky and metallic material is left as a kind of "space rubble."

Comet Kohoutek, discovered in March 1973 and visible to the unaided eye in December 1973 and January 1974, is probably the most-studied comet in history. By fortunate coincidence, it was at its brightest during the Skylab III space mission.

Assuming that comets came into existence with the rest of the solar system, it may seem puzzling that there are any comets left, since a comet loses some of its frozen material each time it passes relatively near the sun. It has been suggested that the solar system is surrounded by a cloud of billions of comets extending as far as 100,000 astronomical units from the sun. A combination of gravitational circumstances sometimes deflects a comet into a new course which sends it close to the sun. According to this view, disintegrating comets are replaced by others drawn in from this endless reservoir.

METEORS

On any clear night several objects streak through the sky, leaving momentarily bright trails of light. These phenomena are *meteors*, often inaccurately called "shooting stars." When tiny objects collide with the earth's atmosphere, the heat from the friction causes meteors. The tiny objects are called *meteoroids*. Collisions occur when the earth and meteoroids reach intersections of their orbits at the same time. When one of these objects passes through the atmosphere without being entirely consumed and strikes the earth's surface, it is termed a *meteorite.* Spectacular meteors—sometimes dazzlingly bright—are referred to as *fireballs* or *bolides*.

There are untold billions of meteoroids. Most of them are much smaller than a grain of rice, but some weigh several tons. Few sizeable meteoroids are actually able to reach the surface of the earth to become meteorites; most are consumed between 40 and 100 kilometers above the earth's surface. The risk of injury or damage from meteorites is exceedingly small, although there have been rare instances of such. Something struck a sparsely populated area of Siberia in 1908, toppling hundreds of square kilometers of forest and breaking windows 80 kilometers away. However, this may have been a comet rather than a meteorite. Another great impact occurred in Siberia in 1947. The ancient Barringer Crater in Arizona is a kilometer wide and 200 meters deep. Although few meteorites of noticeable size strike the earth, there is a continual fall of meteoritic dust; the mass of the earth is probably increased by more than 1,000 tons daily by these *micrometeorites*. (See Figure 2.32.)

FIGURE
2.32

Stone meteorite weighing 338 kilograms from Paragould, Arkansas. (Yerkes Observatory photograph.)

Although meteors can be seen sporadically on any clear night, there are predictable occasions when meteor showers occur. These showers take place when the earth passes through an intersection of its orbit with the orbit of a comet. The predictability of this phenomenon suggests that showers of meteors result from the "rubble" left behind by comets. The sporadic meteors may be fragments from asteroids. Meteors in showers appear to come from a rather small area of the sky, the *radiant*. Showers are generally named for the constellation in which the radiant appears; for example, the Perseid shower occurs about August 12.

Our understanding of meteors and meteorites has grown slowly. Myths and legends developed about these "stones from heaven." Meteorites are avidly sought for scientific study, and about 2,000 have been found. In recent years, photographic "patrols" by a network of cameras have produced meteor photographs taken simultaneously at two or three locations. This has helped in determining altitude, speed, orbit, and approximate point of impact of the meteorites.

In terms of composition, meteorites are classifield as stones, irons, and stony-irons. Stony meteorites, which are the most common, consist largely of silicates. Iron meteorites are composed largely of mixtures of iron and nickel. The stony-irons are mixtures of all these materials. It appears that most meteorites are from the meteors that occur sporadically and not from showers. Thus the meteorites that have been collected are thought to have come from asteroids rather than comets. It is believed that collisions among asteroids have created the fragments we find as meteorites.

ORIGIN OF THE SOLAR SYSTEM

It appears that the entire solar system—the sun and the objects orbiting it—came into existence at about the same time, some 4 to 5 billion years ago. How? Astronomers have been speculating about this with increasing sophistication for a long time. In doing so, they assume that the physical principles evident here and now have applied throughout the universe and all past time. No theory of the origin of the solar system neatly accounts for everything with total and compelling consistency. Many theories have been popular for a time, only to be rejected when they were found to be inconsistent with physical principles.

Several explanations have been rejected because they failed to account adequately for the *angular momentum* of the solar system objects. An object rotating about an axis or revolving around another object possesses angular momentum. The quantity of angular momentum possessed by a body depends upon its mass, the speed of the circular motion, and the distance of the body (or the portion of the body being considered) from the center about which it moves. Angular momentum is conserved in a system of rotating or revolving bodies; it can be transferred from one body to another, but it cannot be created or destroyed. An illustration of this principle is the fact that figure skaters spin more rapidly when they lower their arms and redistribute their mass.

Most explanations of the origin of the solar system fall within two groups of suggestions: planetesimal or tidal hypotheses and nebular hypotheses. The various older tidal hypotheses suggested that the sun and another star passed so close to each other that the material in the objects now orbiting the sun was drawn away from the sun or the passing star by gravitational attraction. The nebular hypotheses propose that the sun and other solar system objects condensed from a cloud of dust and gas through gravitational attraction. Explanations involving condensation are favored at this time. Variations of the condensation hypotheses include suggestions that the sun captured meteoritic material which formed the cores about which material condensed into planets or that the sun encountered dust and gas and captured it in its magnetic field.

CONCLUSION

The earth, rather than being the center of the universe, is one of many objects orbiting an ordinary star. The solar system is rather flat, very spacious, and vastly isolated. The planets, except Pluto, orbit in a region the shape of which has been compared to a phonograph record. It is nearly six billion kilometers from the sun to Pluto, and it is much farther to some of the comets. In a scale model of the solar system, if the sun were represented by a grapefruit, Pluto would have to be represented by something smaller than the head of a pin more than a half kilometer away. On this same scale, the closest star to the sun would be represented by another grapefruit 1,600 kilometers away.

FOR REVIEW, DISCUSSION, OR FURTHER STUDY

1. Which one of the following statements is correct? It takes about one week for the moon to change from
   (a) first quarter to last quarter.
   (b) new moon to last quarter.
   (c) full moon to last quarter.
   (d) new moon to full moon.

2. Is the sunspot cycle an 11-year or a 22-year cycle?

3. By what biological process is the energy of the sun "captured" and made available to sustain life on the earth?

4. Is this a year of unusual solar activity or is it a quiet year on the sun?

5. If one noted the position of the moon against the background of stars at a particular time one night, by about how many degrees will its position have shifted at the same hour on the following night?

6. Sir George Darwin, a son of Charles Darwin, suggested an intriguing but now abandoned explanation for the origin of the moon. What was the gist of his hypothesis?

7. Currently what is the prevailing opinion among scientists regarding life on Mars?

8. What is the name of Saturn's largest satellite? Are the possibilities for life on Saturn considered to be greater or less than the possibilities for life on most solar system objects? Why?

9. What are the Leonids?

10. What, very generally, is the shape and size of the solar system?

SUGGESTED ACTIVITIES

Observing Lunar Surface Features

Very little magnification is required for observation of many of the moon's larger surface features. Ordinary binoculars will reveal some of them, and terrestrial telescopes magnifying 20 or 30 times will reveal many. Telescopes of adequate optical quality can be purchased (often in discount department stores) for 10 to 20 dollars. Moon maps are easily obtained; most libraries will have at least the one that was distributed with the February 1969 issue of *National Geographic*. It is interesting to use moon maps to identify features observed with binoculars or small telescopes.

Observing Earthlight

Look for earthlight on nights immediately following the new moon. In addition to the small, bright crescent, it should be possible to see the entire disk of the moon faintly illuminated.

Observing the Moon Illusion

Observe the rising full moon near the horizon. Note that it appears smaller a few hours later.

Keeping Track of the Solar System

The Old Farmer's Almanac is widely available at newstands. It contains such information as the times at which the sun and moon rise and set on each day of the year, the length of twilight throughout the year, the times and nature of eclipses during the ensuing year, the times of rising and setting of the five visible planets, and the times and altitudes at which a dozen bright stars cross the meridian. In addition, it contains a potpourri of quaint bits of information, many of which are of an astrological nature. The astronomical information can help you to make a great variety of naked-eye observations of solar system objects; it can help you to keep track of the sun, moon, Mercury, Venus, Mars, Jupiter, and Saturn.

SUGGESTED READINGS

Anderson, Don L., "The Interior of the Moon," *Physics Today*, Vol. 27, No. 3 (March 1974), 44.

Anderson, Don L., and Thomas C. Hanks, "Is the Moon Hot or Cold?" *Science*, Vol. 178 (22 December 1972), 1245.

Cameron, A. G. W., "The Origin and Evolution of the Solar System," *Scientific American*, Vol. 233, No. 3 (September 1975), 32.

Cameron, A. G. W., "The Outer Solar System," *Science*, Vol. 180 (18 May 1973), 701.

Carr, Michael H., "The Volcanoes of Mars," *Scientific American*, Vol. 234, No. 1 (January 1976), 32.

Chapman, Clark R., "The Nature of Asteroids," *Scientific American*, Vol. 233, No. 1 (January 1975), 24.

Cruikshank, Dale P., and David Morrison, "The Galilean Satellites of Jupiter," *Scientific American*, Vol. 234, No. 6 (June 1976), 108.

Driscoll, Everly, "A Look Inside the Moon," *Science News*, Vol. 103 (7 April 1973), 228.

Driscoll, Everly, "Mariner's Intriguing Evidence of a Formerly Watery Mars," *Science News*, Vol. 103 (10 March 1973), 156.

Eberhart, Jonathan, "The Planetary Name Game," *Science News*, Vol. 108 (1 November 1975), 282.

"Naming the features of Mars is no picnic—in fact, it's more like a war."

Eberhart, Jonathan, "The Road to Mars," *Science News*, Vol. 108 (2 August 1975), 76.

Gibson, Edward G., "The Sun as Never Seen Before," *National Geographic*, Vol. 146, No. 4 (October 1974), 493.

"Grand Unveiling of the Rocks of Venus," *Science News*, Vol. 108 (1 November 1975), 276.

Grossman, Lawrence, "The Most Primitive Objects in the Solar System," *Scientific American*, Vol. 232, No. 2 (February 1975), 30.

Hammond, Allen L., "Lunar Research: No Agreement on Evolutionary Models," *Science*, Vol. 175 (25 February 1972), 868.

Hammond, Allen L., "Lunar Science: Analyzing the Apollo Legacy," *Science*, Vol. 179 (30 March 1973), 1313.

Hammond, Allen L., "Mars as an Active Planet: The View from Mariner 9," *Science*, Vol. 175 (21 January 1972), 286.

Hammond, Allen L., "The New Mars: Volcanism, Water, and a Debate over Its History," *Science*, Vol. 179 (2 February 1973), 463.

Hammond, Allen L., "Solar Variability: Is the Sun an Inconstant Star?" *Science*, Vol. 191 (19 March 1973), 1159.

Hartmann, William K., "The Smaller Bodies of the Solar System " *Scientific American*, Vol. 233, No. 3 (September 1975), 142.

Howard, Robert, "Recent Solar Research," *Science*, Vol. 177 (29 September 1972), 1157.

Howard, Robert, "The Rotation of the Sun," *Scientific American*, Vol. 232, No. 4 (April 1975), 106.

Hunter, Donald M., "The Outer Planets," *Scientific American*, Vol. 233, No. 3 (September 1975), 130.

Ingersoll, Andrew P., "The Meteorology of Jupiter," *Scientific American*, Vol. 234, No. 3 (March 1976), 46.

Lunar Sample Analysis Planning Team, "Fourth Lunar Science Conference," *Science*, Vol. 181 (17 August 1973), 615.

Metz, William D., "Venus: Radar Maps Show Evidence of Tectonic Activity," *Science*, Vol. 192 (30 April 1976), 454.

Murray, Bruce C., "Mars from Mariner 9," *Scientific American*, Vol. 228, No. 1 (January 1973), 48.

Murray, Bruce C., "Mercury," *Scientific American*, Vol. 233, No. 3 (September 1975), 58.

Page, D. Edgar, "Exploratory Journey out of the Ecliptic Plane," *Science*, Vol. 190 (28 November 1975), 845.

Parker, E. N., "The Sun," *Scientific American*, Vol. 233, No. 3 (September 1975), 42.

Pasachoff, Jay M., "The Solar Corona," *Scientific American*, Vol. 229, No. 4 (October 1973), 68.

Pollack, James B., "Mars," *Scientific American*, Vol. 233, No. 3 (September 1975), 106.

Sagan, Carl, "The Solar System," *Scientific American*, Vol. 233, No. 3 (September 1975), 22.

Science, Vol. 193 (27 August 1976).

This issue contains 13 reports arising from the Viking I mission to Mars.

Siever, Raymond, "The Earth," *Scientific American*, Vol. 233, No. 3 (September 1975), 82.

"Skylab Looks at Earth," *National Geographic*, Vol. 146, No. 4 (October 1974), 470.

Photos from Skylab missions.

"The Strange and Cratered World of Mercury," *Science News*, Vol. 105 (6 April 1974), 220.

Van Allen, James A., "Interplanetary Particles and Fields," *Scientific American*, Vol. 233, No. 3 (September 1975), 160.

Weaver, Kenneth F., "Flight to Venus and Mercury," *National Geographic*, Vol. 147, No. 6 (June 1975), 858.

Weaver, Kenneth F., "Journey to Mars," *National Geographic*, Vol. 143, No. 2 (February 1973), 230.

Weaver, Kenneth F., "Mystery Shrouds the Biggest Planet," *National Geographic*, Vol. 147, No. 2 (February 1975), 285.

Wolfe, John H., "Jupiter," *Scientific American*, Vol. 233, No. 3 (September 1975), 118.

Wood, John A., "The Moon," *Scientific American*, Vol. 233, No. 3 (September 1975), 92.

Young, Andrew and Louise, "Venus," *Scientific American*, Vol. 233, No. 3 (September 1975), 70.

FIGURE
3.1
The Great Galaxy in Andromeda: NGC 224, Messier 31.
(Photograph courtesy of the Hale Observatories.)

THE UNIVERSE:
The Scope and Pattern of the Cosmos

3

We still have not dealt with several questions: Where is the earth in the grand design of the universe? Is our solar system unique? Do objects like the earth orbit other stars? How typical of stars is the sun? Do stars move, or is the celestial sphere really as unchanging as it appears to be? Are the stars randomly distributed throughout an infinite universe, or are there systems of stars? As with most important questions tested through the processes of scientific inquiry, the answers have come rather slowly and in relation to the insightfulness with which they were asked.

The Copernican revolution dispelled the misconception of an earth-centered solar system, but it did not immediately eliminate the concept of a crystalline celestial sphere with the stars attached to its inner surface. On a clear, dark night one can see perhaps 3,000 stars. With the telescope, and especially in time-exposure photographs taken through telescopes, uncounted myriads of stars become visible. Even when viewed with telescopes, they remain just points of light; stars are so distant that they do not present visible disks as solar system objects do. It is a triumph of human intellect that we are able to determine the distance, size, temperature, age, motion, mass, and chemical composition of many of these "points of light." We shall discuss how these properties are determined in later chapters. The only obvious differences one sees among the stars are in their brightness and color. These differences are helpful; the differing brightnesses of stars offer some clues about their distances, and differing colors suggest something about their temperatures (ranging from "red hot" to "white hot").

GROUPING THE STARS

Many stars appear to be grouped. A large number of stars form the *Milky Way*; smaller numbers of stars seem to be arranged in familiar *constellations;* still others form a few smaller *open clusters,* such as the sparkling Pleiades. The constellations are illusory groups in terms of cosmic design; the Milky Way and open clusters are very real. To the unaided eye the Milky Way is a band of light on the celestial sphere, sometimes arching from horizon to horizon. A small telescope reveals that the band of light is produced by the combined light of thousands of faint stars. When astronomers first discovered this fact, they took an immense step toward an understanding of the total cosmic design. We will discuss the Milky Way in some detail later in this chapter, so let us now consider only the various smaller types of groups.

Constellations

Constellations are illusory groups, because often the stars that form them are not in close proximity or physically associated in any unusual way. Stars in constellations form their familiar patterns as an accidental consequence of the direction from which we view them; if the stars were viewed from another part of the universe, the patterns would disappear.

Ancient peoples created the concept of constellations by naming several familiar patterns of stars and incorporating these patterns into their mythology. For instance, Orion, the most magnificent of the constellations, was a giant with an upraised club facing the bull Taurus. Cassiopeia was an Ethiopian queen who boasted of her beauty. Cepheus, her husband, was forced to chain their daughter Andromeda to rocks, where she was at the mercy of a sea monster sent in punishment for Cassiopeia's vanity. Andromeda was saved by Perseus. Early star maps were embellished with fanciful drawings of the subjects of the myths. Whether or not you are interested in mythology, your pleasure in viewing the night sky will be enhanced if you can recognize a few constellations and know when and where to look for them. (See Figure 3.2.)

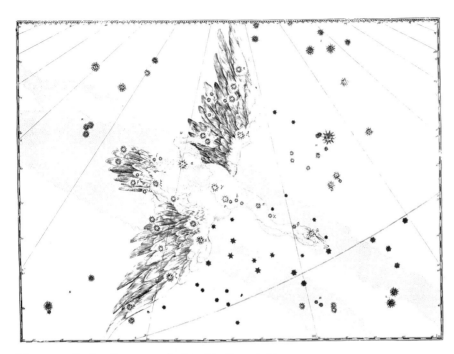

FIGURE
3.2

The constellation Cygnus as depicted in Bayer's *Urano-metria*, 1655. (Yerkes Observatory photograph.)

The ancients identified three or four dozen constellations. Astronomers have now created additional constellations by dividing the entire celestial sphere into 88 constellations or portions. The constellations provide a handy means for locating an object less precisely than by using its right ascension and declination. For instance, saying that Mars is presently in Gemini would enable a person familiar with the constellations to know whether or not Mars was currently visible in the night sky and to know how to find it readily.

Open Clusters

The Pleiades have been cited as an example of an open cluster of stars (Figure 3.3). The Hyades, also located in the constellation Taurus, form another such cluster. The stars in an open cluster are clustered because of common origin and gravitational attraction among themselves. Observations of some clusters have revealed that the member stars are moving on parallel paths at similar speeds.

FIGURE
3.3

The Pleiades, an open cluster in the constellation Taurus.
(Lick Observatory photograph.)

Globular Clusters

Globular clusters of stars are larger and more distant than open clusters. The terms "globular cluster" and "open cluster" are descriptive of the clusters' differing appearances. A telescope is required to reveal the shape and beauty of globular clusters. The brightest globular clusters are barely visible to the unaided eye, because they are several thousand light years away. The typical globular cluster contains several hundred thousand stars. (See Figure 3.4.)

FIGURE
3.4

Globular star cluster in Hercules: NGC 6205, Messier 13.
(Photograph courtesy of the Hale Observatories.)

SOME ATTRIBUTES OF STARS

Apparent Magnitude

Hipparchus, a Greek astronomer, attempted to systematize observations of the brightness of stars more than 2,000 years ago. He classified the stars according to brightness in six groups or magnitudes. The brightest stars were of the first magnitude; those just barely visible were of the sixth magnitude. A star that differs in brightness from another star by one magnitude (for instance, third magnitude compared to fourth magnitude) is actually 2.512 times as bright. A star of the first magnitude is 100 times as bright as a star of the sixth magnitude (2.512^5).

The ancient system is still used, but astronomers can now measure and compare brightness much more accurately than Hipparchus could. The brightness of stars can be measured using *photoelectric photometers,* sensitive light meters attached to telescopes. Differences as small as 0.01 of a magnitude can be detected. For instance, the star Spica has a magnitude of 1.21, and the star Antares has a magnitude of 1.22. Stars and other celestial objects brighter than a first magnitude star have negative magnitudes; the magnitudes of Sirius, the full moon, and the sun are -1.4, -12.6, and -26.9, respectively. The faintest objects visible in the largest telescopes have magnitudes of about 23. Because these magnitudes are all measurements of the brightness of a star as seen from the earth, they are called apparent magnitudes.

Absolute Magnitude

The apparent magnitude of a star results from both its true brightness and its distance. A star that is actually brighter than another may appear fainter if it is at a greater distance. If the distance to a star is known, it is possible to calculate the apparent magnitude it would have at a different distance. Astronomers find it helpful to compare the apparent magnitudes of stars at a uniform distance of 10 parsecs; these magnitudes are referred to as *absolute magnitudes.* One *parsec* is the distance at which the radius of the earth's orbit around the sun would subtend an angle of one second; it is also the distance at which a star would have a parallax of one second. Ten parsecs is a great distance (about 300 trillion kilometers), but most stars are more distant than 10 parsecs. The sun is the only star closer than one parsec. Although the sun's apparent magnitude is about -27, its absolute magnitude is only 4.8. At a distance of ten parsecs, the sun would not be a bright star. (See Figure 3.5.)

Designations

There are far too many stars for all of them to be given proper names, although the brightest stars have been named since ancient times. Many

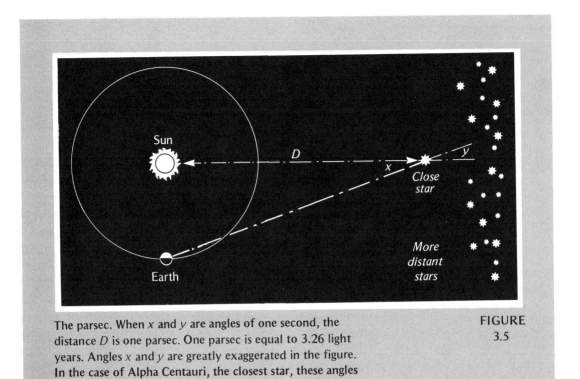

The parsec. When x and y are angles of one second, the distance D is one parsec. One parsec is equal to 3.26 light years. Angles x and y are greatly exaggerated in the figure. In the case of Alpha Centauri, the closest star, these angles equal 0.751 second.

FIGURE 3.5

of these names were derived from Greek and Arabic names and expressions. Sirius comes from a Greek word meaning sparkling or scorching; Algol is derived from an Arabic expression meaning the demon's head. Using Ptolemy's old system of nomenclature, Sirius is designated as Alpha Canis Majoris and Algol as Beta Persei. In this system, the letters of the Greek alphabet are assigned to stars in a constellation, usually in order of brightness. Thus, Sirius is the brightest star in Canis Major and Algol is the second brightest star in Perseus. The names of astronomers have become associated with some stars, such as Barnard's Star. However, most stars are simply designated by the numbers assigned to them in particular catalogs. For instance, HD6309 indicates star number 6309 in the Henry Draper catalog of 225,300 stars. In addition to general catalogs of stars and catalogs of particular kinds of stars, there are also catalogs of nebulae and galaxies, objects that will be discussed presently. A famous catalog of this type is Charles Messier's list of 103 "bright objects" published in 1784. M31 indicates the thirty-first object in Messier's catalog.

Distances

It is difficult to comprehend the immensity of stellar distances. The kilometer is an inappropriately small unit for measuring distances beyond our solar system. Two units for measuring great distances have been mentioned—the astronomical unit and the parsec. Another unit used more often than either of these in measuring stellar distances is the light year; a *light year* is the distance that light can travel in one year. Since light travels at about 300,000 kilometers per second, a light year is about $9\frac{1}{2}$ trillion kilometers. Sirius, the nearest star visible from most of North America, is at a distance of about $8\frac{1}{2}$ light years, whereas the light from the sun, our own star, requires only about eight minutes to reach the earth. Two stars in the sky over the Southern Hemisphere are closer than Sirius; Proxima Centauri and Alpha Centauri are each at a distance of about four light years. There are only a dozen or so stars within 10 light years of the solar system; many of the stars we see are several hundred light years away. We are looking into the past, then, when we view the stars; the light by which we view Polaris, the North Star, has been traveling for about 700 years.

The method of measuring distances by parallax was described in Chapter 1; additional methods will be discussed later. It was very difficult to detect stellar parallax, because the parallactic displacements are so small, even for the closest stars. Many astronomers tried for many years, but the parallactic displacement of a star was not observed until 1838. Efforts to observe parallax resulted in other important discoveries, including the discovery of *visual binary stars.*

Proper Motion

Another discovery that preceded that of the parallactic displacement of stars was the discovery that the stars actually move and are not fixed to a celestial sphere. In 1718, Edmund Halley made careful comparisons of the current positions of several stars and their positions as given by Ptolemy several centuries earlier. Their positions had apparently changed. He concluded that the changes were too significant to be written off as errors in Ptolemy's work; he decided that the stars had actually moved.

The movement of a star revealed by changes in its right ascension and declination is called *proper motion*. Astronomical photography has facilitated the detection and study of proper motion. Two photographs of the same region of the sky taken 20 or more years apart may be superposed to detect changes in relative position. Proper motions, difficult as they are to detect, are actually greater than parallactic displacements.

The proper motion of a star depends upon the distance to the star. If the distance and proper motion of a star are known, it is possible to calculate the velocity of the star at right angles to the line of sight; this is its *tangential velocity*. The speed with which an object is moving toward or away from the earth is its *radial velocity*. If tangential and radial velocities are known, it is possible to calculate the star's true velocity in relation to the solar system. A method for determining radial velocity will be discussed later.

SPECIAL TYPES OF STARS

Visual Binary Stars

Sir William Herschel, discoverer of Uranus, was among those who tried hard to observe stellar parallax. He observed some "pairs" of stars that he believed to be two stars at greatly different distances but in the same line of sight. He hoped to see a parallactic displacement of the nearer star in relation to the more distant star. Instead of observing this, he made the important discovery that there actually are many pairs of stars in which one stars revolves around the other. More accurately, they each revolve around a common center of gravity, but the apparent effect (as with the moon and the earth) is that one revolves around the other. Herschel was not the first astronomer to see that some stars are doubles, but earlier observers failed to understand that such stars are literally paired and physically associated. Several hundred pairs of such double stars have been discovered; their periods of revolution range from about two years to hundreds of years.

Variable Stars

Many stars vary in brightness, often in a regular manner. The above discussion of binary stars suggests an explanation of some variable stars.

Sometimes the orbital plane of a pair of visual binary stars is in our line of sight and one star eclipses the other, either partially or totally. Such stars are *eclipsing binaries*. Other stars, called *intrinsic variables*, change in brightness because of their own structure. Perhaps such stars are alternately expanding and contracting, as outward forces resulting from heating and inward forces from gravitation become alternately greater and lesser.

Several thousand intrinsic variables have been discovered. The light curve, which is a plot of magnitude against time, helps to reveal the characteristics of a variable. Depending upon their characteristics, variables are categorized in several groups. The periods between maximum brightness range from about an hour to a year or more. Some variables differ in brightness by as much as five magnitudes at various times; the brightness of other variables may change by only 0.1 magnitude.

The *Cepheid variables* are a particularly interesting group of intrinsic variables. The group gets its name from the first such star to be discovered, Delta Cephei. These stars provide a means of determining some stellar distances, because there is a correlation between the periods of Cepheids and their absolute magnitudes. From its period, one can therefore know the absolute magnitude of a Cepheid. Knowing its absolute magnitude and apparent magnitude, one can calculate its distance.

The *RR Lyrae variables* are another type of variable for which distances can be determined rather readily. The RR Lyrae variables are bright variable stars with periods of less than one day. It has been concluded that all of them have an average absolute magnitude of about 0.5. Just as with a Cepheid, the distance of an RR Lyrae variable can be calculated by comparing its absolute and apparent magnitudes.

Spectroscopic Binaries

It was mentioned in the last chapter that the spectroscope is an instrument for studying light emitted by luminous bodies or absorbed by gases. Later we will discuss the physical principles that make it one of the most useful devices in astronomical research. The spectroscope permits the astronomer to determine whether a source of light is approaching or receding and how fast it is moving. This is possible because the wavelengths of light are "crowded together" and shortened if a star is speeding toward the earth; and if it is moving away, the wavelengths are lengthened. The spectroscope has revealed many "double" stars that are not visual binaries; such pairs are *spectroscopic binaries*. In such cases, the spectroscope reveals that light that appears to be coming from a single star is really coming from a pair of revolving stars; it reveals that each star in the pair is alternately approaching the earth and receding from it.

Mizar, a second magnitude star at the bend of the Big Dipper's handle, was both the first visual binary and the first spectroscopic binary to be discovered. In 1650, telescopic observation revealed that Mizar, which appeared to be a single star, was two stars. In 1889, spectroscopic analysis revealed that the brighter star of the pair was actually another pair. Mizar is a "triple star."

The stars in the visual and spectroscopic binaries that constitute Mizar should not be confused with Alcor, the "naked-eye companion" of Mizar. Alcor is a pretty little fourth-magnitude star separated from Mizar by about eleven minutes of arc. The ability to see Alcor is a famous old challenge to the sharpness of vision, although it really isn't much of a challenge.

Novae and Supernovae

In 1054 there occurred a cosmic explosion that dwarfs all earthly blasts, including Krakatoa. Where there had apparently been no star, a star brighter than all others suddenly appeared. Although the star faded from view again in a few months, the telescope still reveals a huge, rapidly expanding cloud of gas left from this explosion of nearly 1,000 years ago. This is the Crab Nebula in the constellation Taurus (See Figure 3.6.) The star that became visible was a *supernova*. Other supernovae occurred in 1572 and 1604. In this context, the term *nova* means *new star*. Supernovae are not literally new; rather, they are explosions of stars that existed previously. The explosion involved in a supernova does not destroy the star completely; apparently only the outer portion of a star is blown away. *Novae* look like similar but less violent events. Originally they were thought to be such, but now they are believed to have a different kind of origin. They are thought to be involved somehow in the evolution of binary stars.

Pulsars

A puzzling new kind of variable star was discovered by radio astronomers in 1968; many more of these objects were found over the next

FIGURE
3.6

Crab Nebula in the constellation Taurus, remains of a
supernova of A.D. 1054: NGC 1952, Messier 1.
(Photograph courtesy of the Hale Observatories.)

few years. These objects emit bursts of radio waves at very short and very
regular intervals—intervals ranging from a few hundredths of a second
to a few seconds. So regular are these pulses that at first some suggested
they might be signals from intelligent creatures. The objects were given
the name *pulsar,* an acronym for *pulsating radio source.* It was not until
a year after the discovery of these radio sources that a pulsar was
identified visually—a star in the Crab Nebula in Taurus that had been
known for years. It is believed to be the remains of the star that
produced the supernova of 1054 A.D. Pulsars are thought to be immensely dense, rapidly rotating objects smaller than the earth. They are
believed to be neutron stars formed when stars "collapse" under pressure and their protons and electrons unite to form neutrons. A
teaspoonful of material from such a star would weigh thousands of tons.

NEBULAE

Planetary Nebulae

A *planetary nebula* is a shell of gas surrounding a central star. Planetary nebulae may result from stellar explosions in which clouds of gas are expelled by the central star. They are called planetary nebulae because many of them have the greenish color of Neptune or Uranus and appear as disks in telescopes. (See Figure 3.7.)

FIGURE 3.7

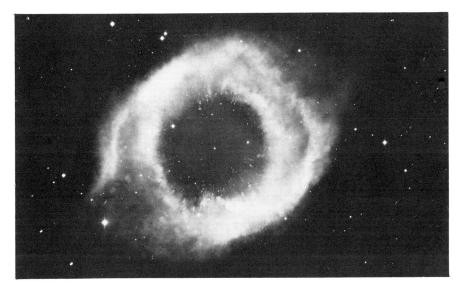

Planetary nebula in Aquarius: NGC 7293. (Photograph courtesy of the Hale Observatories.)

Diffuse Nebulae

Out among the stars there are huge, irregular clouds of dust and gas known as *diffuse nebulae*. An example of a diffuse nebula is the beautiful Great Nebula in Orion, which has a diameter of more than 25 light years. They are very tenuous—as rarefied as the best laboratory vacuums. Although they look impressive on film, this is a result of their size, not their density. Some are called *bright nebulae*, because they are luminous. This luminosity results from the fact that they reflect the light of nearby stars or the fact that excitation by nearby stars causes their atoms to emit a "cold" light. The light of bright nebulae, then, unlike that from stars, does not result from hydrogen bomb-like thermonuclear processes. *Dark nebulae* are nonluminous; their shapes are seen silhouetted against luminous objects beyond them. (See Figures 3.8 and 3.9.)

FIGURE
3.8

The Orion Nebula, a bright nebula: NGC 1976, Messier 42.
(Lick Observatory photograph.)

FIGURE
3.9

The Horsehead Nebula, a dark nebula. (Photograph
courtesy of the Hale Observatories.)

INTERSTELLAR MATTER

The vast space that separates the stars and the diffuse nebulae is not a
total void. The high vacuum of space is not a complete vacuum. Radio
astronomers have known for years that interstellar space is permeated
with atoms of hydrogen radiating energy with a wavelength of 21
centimeters. Recent advances in radio astronomy have revealed the
presence of more than 30 other substances in interstellar space, including ammonia, water, formaldehyde, carbon monoxide, hydrogen cyanide,
and cyanogen. These discoveries raise difficult new questions about the
origin of these substances. Interstellar chemistry has become an important new area of astronomical investigation.

OTHER PLANETARY SYSTEMS

Do planets revolve around other stars, or is the sun with its solar system
unique? This old and highly intriguing question remains only poorly

answered. No one has been able to see a planet orbiting another star. The vast distances to the stars make such direct observation impossible with present-day telescopes, even if planetary systems exist. However, evidence of the kind that suggested the existence of Neptune before its discovery seems to indicate the presence of planets orbiting a few of the closer stars. Barnard's Star is the second closest star to the earth, and it has the largest proper motion yet discovered. Analysis of thousands of photographs of Barnard's Star accumulated during more than a quarter century reveals a slight oscillation in its proper motion. The difficulty of this kind of analysis is suggested by the fact that in a century the proper motion of Barnard's Star is less than thirty minutes of arc. Thirty minutes of arc is the angular diameter of the full moon. The oscillation in the motion of Barnard's Star could be explained by hypothesizing the presence of one or two planets of roughly the mass of Jupiter.

At present, then, evidence for the existence of planetary systems is rather meager. Astronomers believe it likely, however, that a vast number of stars have planetary systems, since they believe that the conditions that led to the formation of the sun and its planetary system are rather ordinary.

GALAXIES

The Milky Way

Do the stars, star clusters, and nebulae combine to make a larger unit in the cosmic design, or do they occur randomly throughout the universe? They constitute a larger unit—the Milky Way Galaxy. To the unaided eye, the Milky Way is a hazy band of light that encircles the earth and roughly divides the sky into two hemispheres. The telescope reveals the milky light to be the combined light of myriads of stars at great distances. The telescope also reveals that dark and bright nebulae are more common in the plane of the Milky Way than elsewhere. The earth, the solar system, and all of the stars that we can see without a telescope are within the Milky Way Galaxy. (See Figure 3.10.)

Astronomers have had a difficult time measuring and mapping the gigantic Galaxy from our little 13,000 kilometer ball of rock. The work

FIGURE
3.10

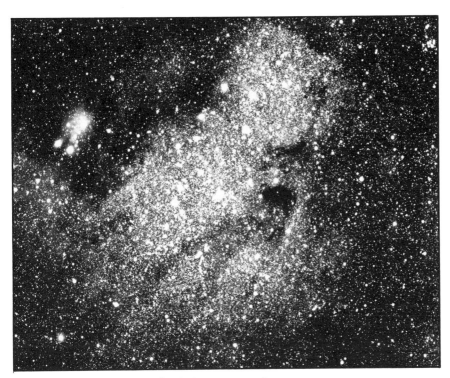

A portion of the Milky Way. (Lick Observatory photograph.)

still goes on, although the general design of our Galaxy has been fairly
well understood for about a half-century. If we could leave the Galaxy
and view it from a distance, it would look like Figure 3.11 or 3.12,
depending upon the angle from which it was viewed. These figures are,
of course, photographs of other galaxies; despite the vastness of the
Milky Way Galaxy, it is not the entire universe. Our Galaxy is a colossal
assemblage of stars and nebulae in a volume of space that is shaped
somewhat like a reading lens. It is about 100,000 light years across and
about 10,000 light years thick in the middle. It contains more than 100
billion stars. Most of the stars are located in the thick central region. It
is termed a *spiral galaxy*, because some stars and nebular material trail
in thin arms that spiral about the central region. The entire arrangement
is "incased" in a sphere of globular clusters; the globular clusters do not
lie in the plane of the Galaxy. The sun is about 30,000 light years from
the center, which lies in the direction of the constellation Sagittarius.

FIGURE
3.11

A spiral galaxy in Coma Berenices seen edge-on: NGC 4564.
(Photograph courtesy of the Hale Observatories.)

FIGURE
3.12

The Whirlpool Galaxy, a spiral galaxy in Canes Venatici:
NGC 5194, Messier 51. (Photograph courtesy of the
Hale Observatories.)

The spiral shape of the Galaxy suggests that the stars that comprise it are rotating about the center of the galaxy. They are. The time required for the sun to complete a galactic revolution is about 200,000,000 years, which is sometimes referred to as the *galactic year*.

William Herschel made the first careful attempt to determine whether or not all of the known stars comprised an orderly system. In the 1780s he made counts of the stars in nearly 700 carefully chosen regions of the sky. He concluded that the sun was near the center of a system of stars and that this system was shaped like the double-convex lens of a reading glass. He thought it to be about five times as broad as it was thick, and he believed the Milky Way to be in the plane of the broad portion. He was unable to assign dimensions to his model of the stellar system.

Herschel's star-count approach to the problem was refined by others—particularly Jacobus Kapteyn. Kapteyn, who worked in the early years of this century, was able to reach some conclusions about the size of the system; he believed it to be about 10,000 light years across and about 4,000 light years thick across the middle. A different approach to the problem was taken by Harlow Shapley about 1917. Shapley plotted the positions of globular clusters rather than single stars. He concluded that the globular clusters occupied a nearly spherical volume of space around the flattened array of stars and nebulae in the plane of the Milky Way. Thus he decided that the sun was far from the center of the Milky Way. Shapley was able to observe RR Lyrae or cluster variables in some globular clusters; because of the uniformity in absolute magnitude of these stars, he was able to estimate their distances. He then concluded that the diameter of the galaxy was about 200,000 light years and its thickness about 30,000 light years. Subsequent refinements in measurements have scaled these dimensions down to 100,000 light years by 10,000 light years. The task of mapping the details of the Milky Way Galaxy is far from completed. Radio astronomy observations of the hydrogen gas and other interstellar material are helping greatly to sharpen our picture of the trailing arms in our spiraling Galaxy.

Galaxies Beyond the Milky Way

The Milky Way Galaxy was generally considered to be the entire universe at the beginning of this century. This insular and inadequate concept of the universe might be expected, for 100 billion stars scattered through 100,000 light years of space is enough of a universe in itself to stagger the mind. Furthermore, telescopes of the day could not penetrate clearly into the incredible distances stretching off in all directions from our Galaxy.

Shapley's views of our own Galaxy were challenged by H. D. Curtis. The two met in a great debate at the National Academy of Sciences in Washington on April 26, 1920. In addition to debating the size and structure of our Galaxy, they debated the possibility of other galaxies beyond our Milky Way. Some other spiral galaxies had been seen indistinctly, but they were generally regarded to be nebulae in our own Galaxy. Curtis contended that the *spiral nebulae* were actually other galaxies at enormous distances. He had compared the brightness of novae in the Andromeda spiral with the brightness of novae in the Milky Way and had concluded that the Andromeda spiral was at a distance of 500,000 light years. (We believe now that the great spiral Andromeda, M31, is at a distance of about 2,000,000 light years.) Neither viewpoint could be proven until better telescopes became available. In 1924, Edwin Hubble was able to photograph individual stars in the arms of the Andromeda spiral using the new 100-inch telescope on Mount Wilson. (See Figure 3.1.) Some of these were Cepheid variables, and Hubble concluded that they were even more remote than Curtis had estimated. The matter was settled; it was clear that the Milky Way Galaxy was only one of many galaxies.

Since Hubble's day, astronomers have continued to peer ever deeper into space, discovering countless galaxies which appear to spread away endlessly in all directions. About 40% of the innumerable galaxies are spiral galaxies. Others are *irregular galaxies* and *elliptical galaxies.* The elliptical galaxies seem to be essentially without the dust and gas present in spiral and irregular galaxies. (See Figures 3.13 and 3.14.)

We can now see several billions of light years out into the universe. When we look 5 billion light years into space, we are also looking 5 billion years into the past. Have we looked almost to the end of space? Have we looked back almost to the beginning of time?

Clusters of Galaxies

Inquiry into nature has revealed that a multitude of stars, star clusters, and nebulae together constitute the gigantic Milky Way Galaxy and that

FIGURE
3.13

The large Magellanic Cloud, an irregular galaxy. (Lick
Observatory photograph.)

FIGURE
3.14

An elliptical galaxy: NGC 205. (Photograph courtesy of the
Hale Observatories.)

this galaxy is only one among galaxies as numerous as the stars in the Milky Way. Isn't it appropriate to inquire whether the galaxies are grouped in any way or whether they are distributed uniformly throughout the universe?

Counts of galaxies in particular regions of space reveal that they are indeed grouped; they are grouped into clusters. Some clusters contain as few as two galaxies, whereas some clusters contain hundreds. The Milky Way Galaxy is one of about twenty known members of the *Local Group* of galaxies. The Milky Way Galaxy and the Andromeda Galaxy are the largest members of the Local Group. (See Figure 3.15.)

FIGURE
3.15

Cluster of galaxies in Hercules. (Photograph courtesy of the Hale Observatories.)

Inquiry into the realm of the galaxies has only begun, but already there appears to be some evidence supporting the existence of clusters of clusters.

THE EXPANDING UNIVERSE

The other galaxies seem to be moving away from us in all directions with ever-increasing speeds as distances lengthen. Does this mean that the earth is at the center of the universe, as Ptolemy and other ancient thinkers believed? Since the earth is not even at the center of the solar system let alone the center of the Milky Way, surely it is not at the center of an expanding universe! No; this same deceptive view would be seen from any galaxy. With all intergalactic distances increasing and with galaxies at greater distances from the center moving at greater speeds, the view from any galaxy would seem to place the observer at the center of the universe. The recession and speed of the other galaxies is determined spectroscopically from the lengthening of wavelengths of light reaching us from the galaxies; this technique was mentioned in connection with spectroscopic binaries. The most distant visible galaxies are hurtling outward into the void at speeds of more than 80,000 kilometers per second.

If the universe is expanding, it might seem that at some time in the distant past all of the material in the universe must have been together. The *"big-bang" theory* pictures the universe beginning with a stupendous explosion of an extremely dense central mass about 13 billion years ago.

Cosmology is the branch of astronomy particularly concerned with the creation and total structure of the universe. The "big-bang" concept is only one of several cosmological theories. A *steady-state universe* was a popular concept a few years ago. This pictured the continual creation of matter to replace that moving beyond our view in an expanding universe. Some suggest that the universe pulsates in alternate periods of expansion and contraction. Cosmologists deal with such disconcerting problems as whether space is finite or infinite and how it is shaped.

QUASARS

A cosmological enigma was discovered in the early 1960s; this is the class of objects referred to as *quasi-stellar radio sources* or *quasars*. Radio astronomers found powerful emissions coming from several unseen sources of very small angular diameter. Efforts were made to photograph these objects using large optical telescopes and measurements of right ascension and declination provided by radio astronomers. The radio waves were found to be coming from blue-colored objects fainter in brightness than the twelfth magnitude. The lengthening of their light waves (the "red shift") is the greatest ever observed. This seems to indicate they are the most remote and most rapidly receding objects ever seen; they seem to be 10 to 15 billion light years away and moving at speeds of more than 160,000 kilometers per second. Apparently they are much smaller than most galaxies and yet they emit many times more radiation than the most luminous galaxy. They seem to be emitting more energy than can be accounted for by any known process. What are quasars? This is one of the most stimulating questions facing astronomers today. Some think that they are an unusual type of galaxy which existed in the dim past. (Something viewed across 10 billion light years of space is seen as it was 10 billion years ago.) Others think that quasars are within or close to our Milky Way Galaxy. According to this view, the lengthening of their light waves results from intense gravitational influences or from explosions in our or nearby galaxies. Most people believe that they are at great distances. Whatever and wherever they are, most astronomers believe that successful inquiry into the nature of quasars will bring important new perspectives on the cosmic design.

CONCLUSION

The fixity of the visible stars is illusory. Stars are viewed from such great distances that speeds of several miles per second result in such small changes in apparent position that the stars appear to be affixed to a turning sphere. The few thousand visible stars are only our sun's neighbors among billions of unseen stars in our Milky Way Galaxy. Our Galaxy is one of millions of such gigantic systems in an expanding universe of galaxies. The sun is a very ordinary star, although stars differ in several ways. Evidence of planetary systems around other stars is scanty, but the sun's ordinary nature suggests that the solar system is not unique.

FOR REVIEW, DISCUSSION, OR FURTHER STUDY

1. What is the grand design of the universe and where is the earth's place in it?

2. Is there evidence of the existence of planets orbiting stars other than the sun?

3. Name the brightest star in each of the following constellations: Cygnus, Aquila, Gemini, Auriga, Orion, Canis Major, Canis Minor, Boötes, Scorpius, and Leo.

4. In what season would an observer in mid-America see Orion in the south in the evening? Scorpius?

5. How many times brighter is a star of apparent magnitude 2 than one of apparent magnitude 4?

6. Are Alcor and Mizar actually the component stars of a binary?

7. What is meant by proper motion, radial velocity, and tangential velocity?

8. How are the distances of Cepheid and RR Lyrae variables determined?

9. Why are quasars puzzling?

10. What is the arrangement of the globular clusters associated with our Galaxy in relation to the plane of the spiral?

SUGGESTED ACTIVITIES

Astrophotography with a 35mm Camera

It is remarkably easy to take striking and informative astronomical photographs, some of which reveal things unseen by the naked eye. All that is required is a 35 mm camera, a tripod, and a fast lens (an f/1.4 works very well). High-speed Ektachrome or comparably fast color film should be used. Constellation photographs can easily be taken with high-speed Ektachrome film (processed for ASA 400), 15-second exposures, and an f/1.4 lens. With exposures of more than 30 seconds, the rotation of the earth will cause the stars to form trails on the film rather than points. The differing colors of stars are much more evident in photographs than to the naked eye. The Orion Nebula, which is difficult to see with the unaided eye, is easily seen as a red spot in such photographs. It is revealing to photograph Venus or Mars several times over a period of several weeks, noting the changing positions of these planets against the background stars. The lack of a fast lens need not deter one from taking astrophotographs. With slower lenses (or, for that matter, with fast lenses), it is interesting to photograph star trails in exposures of several minutes. A five-minute exposure of Orion, for instance, provides a striking photograph of varicolored star trails, unmistakably those of Orion. Long-exposure photographs centered on Polaris are also interesting.

SUGGESTED READINGS

Burbridge, Geoffrey and Margaret, "Stellar Populations," *Scientific American*, Vol. 199, No. 5 (November 1958), 44.

Hewish, Anthony, "Pulsars," *Scientific American*, Vol. 219, No. 4 (October 1968), 25.

Hewish, Anthony, "Pulsars and High Density Physics," *Science*, Vol. 188 (13 June 1975), 1079.

   A lecture delivered by Professor Hewish when he received the Nobel Prize in Physics, a prize that he shared with Professor Ryle.

Holcomb, Robert W., "Galaxies and Quasars: Puzzling Observations and Bizarre Theories," *Science*, Vol. 167 (20 March 1970), 1601.

Metz, William D., "Quasars Flare Sharply: Explaining the Energy Gets Harder," *Science*, Vol. 189 (11 July 1975), 129.

Miller, Joseph S., "The Structure of Emission Nebulas," *Scientific American*, Vol. 231, No. 4 (October 1974), 34.

Oort, J. H., "Galaxies and the Universe," *Science*, Vol. 170 (25 December 1970), 1363.

Percy, John R., "Pulsating Stars," *Scientific American*, Vol. 232, No. 6 (June 1975), 66.

Rubin, Vera C., "The Dynamics of the Andromeda Nebula," *Scientific American*, Vol. 228, No. 6 (June 1973), 30.

Sanders, R. H., and G. T. Wrixon, "The Center of the Galaxy," *Scientific American*, Vol. 230, No. 4 (April 1974), 66.

Stephenson, F. Richard, and David H. Clark, "Historical Supernovas," *Scientific American*, Vol. 234, No. 6 (June 1976), 100.

Thomsen, Dietrick E., "Mystery of the Cosmos," *Science News*, Vol. 103 (20 January 1973), 46.

   A brief article about quasars.

Turner, Barry E., "Interstellar Molecules," *Scientific American*, Vol. 228, No. 3 (March 1973), 51.

Wade, Nicholas, "Discovery of Pulsars: A Graduate Student's Story," *Science*, Vol. 189 (1 August 1975), 358.

Wick, Gerald L., "Interstellar Molecules: Chemicals in the Sky," *Science*, Vol. 170 (9 October 1970), 149.

FIGURE 4.1

The launch of the Apollo 17 space vehicle, December 7, 1972 at 12:33 a.m. EST, carrying astronauts Cernan, Evans, and Schmidt on the sixth and last lunar landing mission in the Apollo program. (NASA photograph.)

MOTION
AND GRAVITATION:
Celestial Mechanics

4

Does it seem obvious that an object at rest will remain at rest unless some force sets it in motion? Does it seem equally obvious that an object moving steadily in a straight line will continue to do so until a force stops it or changes its direction or its speed? To most people these concepts now seem obvious, but actually we have understood motion and its causes for a relatively short time. Many feel that modern physical science began when Galileo arrived at his *principle of inertia,* which was restated by Newton as his *first law of motion: An object will remain at rest or in uniform motion in a straight line unless acted upon by an unbalanced force.*

ARISTOTELIAN VIEW OF MOTION

For many centuries prior to Galileo, the writings of Aristotle (384?–322? B.C.) influenced science profoundly. Aristotle held that all earthly matter was made up of four elements—earth, water, air, and fire. The highest place in nature was held by fire; beneath it, in descending order, were air, water, and earth. He believed that most motion was the result of these elements trying to return to their natural places. Aristotle explained that the fall of a stone to the bottom of a lake was the result of the element earth in the stone trying to reach its natural place beneath air and water. Gases, containing much of the element air, would naturally bubble to the top of the water, as the elements air and water sought their natural places. Hot air, containing much of the element fire, would naturally rise above cold air. Aristotle believed that forces (pushes or pulls) could cause unnatural *violent motions.* Thus, an object composed chiefly of the element earth could be hurled upward through air. Objects set in violent motion moved naturally again after the forces ceased to be applied. What then was the explanation of the seemingly anomalous fact that a hurled object continues to move upward after it is released? It was believed that the moving object parted the air. The air then moved to the rear of the object to fill a vacuum left by the movement of the object. The air pushed the object forward as it filled the vacuum. Aristotle believed that heavy objects (those containing more of the element Earth) fell faster than light objects. Celestial objects were believed to be composed of an extraordinary fifth element, the *quintessence.* The natural motion of celestial bodies was to encircle the earth, the center of the universe.

GALILEO AND FALLING BODIES

Galileo studied the motions of falling bodies and challenged Aristotle's explanation. According to Aristotle, if two balls weighing 10 kilograms and 1 kilogram, respectively, were dropped from a height of 10 meters, the 10-kilogram ball would strike the ground when the lighter ball had fallen only one meter. Galileo demonstrated the correctness of his own hypothesis that they would reach the ground simultaneously. Actually he found that the heavier object reached the ground slightly sooner. However, after the invention of the air pump it became evident that, in the absence of air resistance, objects of different weights fall together (see Figure 4.2).

Galileo found that freely falling objects gain speed in a uniform manner. In order to determine this, he devised a clever experimental technique involving inclined planes. Timing devices were inadequate for use in experiments with falling objects. He reasoned that whatever caused balls to fall freely would also cause them to roll down inclines, but more slowly. These slower speeds made it possible for him to measure acceleration.

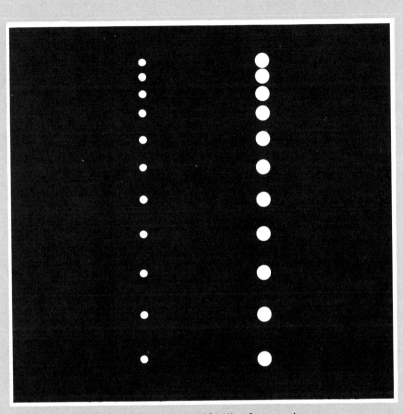

FIGURE
4.2

The motions of freely falling objects. If balls of unequal
weight are released simultaneously and photographed with
stroboscopic light, they are seen to fall at the same rate.
The balls are always at the same height above the ground.

Further discussion of these matters requires careful definition of
the terms *speed, velocity,* and *acceleration. Speed* is rate of motion or
change of position; 30 kilometers per hour is a speed. *Velocity* is speed
in a specific direction: 30 kilometers per hour southward. *Acceleration*
is rate of change of velocity. For instance, if an accelerating object is
increasing its speed by five meters per second during each second, its
acceleration is 5 meters per second per second, or 5 m/sec² . Since velocity
connotes both speed and direction, a change in direction of a

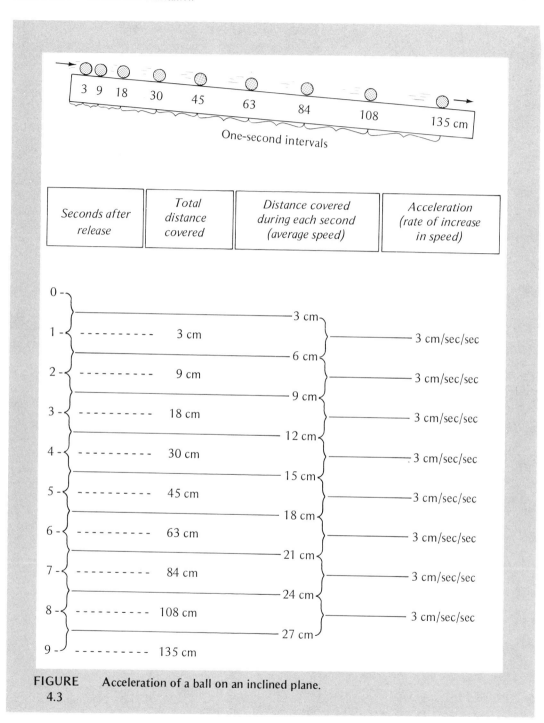

FIGURE   Acceleration of a ball on an inclined plane.
4.3

moving object—even without change in speed—is viewed as acceleration. This view of acceleration as a change in the direction of motion will be helpful in considering motions of celestial objects.

Galileo found that a ball rolling down an inclined plane is accelerated uniformly in the manner illustrated in Figure 4.3. He found, of course, that the steeper the grade, the greater the acceleration. He also

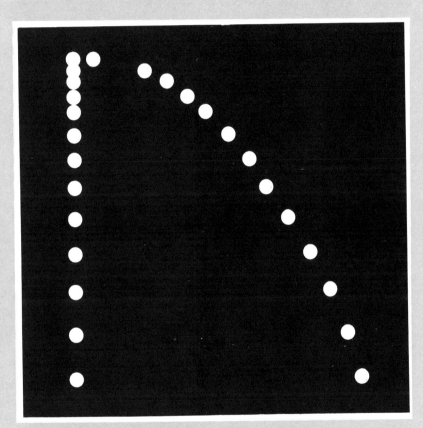

FIGURE
4.4

Forward motion and the acceleration of freely falling bodies. If one ball is dropped vertically and another released simultaneously is given a horizontal motion, they fall at the same rate. If they are photographed with stroboscopic light, they are always seen to be at the same height from the ground.

found that the weight of the ball made no difference. Although he was unable to measure the acceleration during free fall (a perpendicular "incline"), he concluded that it too would be uniform. Now we know that the acceleration of objects falling through a vacuum near the earth's surface is 9.8 m/sec² or 32 ft/sec². Galileo found that the acceleration of a freely falling object is not affected by any additional forward motion it might have. If a bullet is dropped from the height of a gun's muzzle at the same instant that one is fired parallel to level ground, they strike the ground at the same time (Figure 4.4).

It should be noted that Galileo did not try to explain the *cause* of the motion of falling bodies; he did not deal with the force that accelerated them. Rather, he described the nature of the motion. Forty-five years after Galileo's death, Newton dealt with the force (gravitation) and showed that it caused both celestial and earthly motions.

KEPLER'S LAWS OF PLANETARY MOTION

Kepler's First Law

While Galileo was describing terrestrial motions, Kepler was describing celestial motions with new clarity. Kepler's development of the laws of planetary motion was an extremely significant intellectual achievement.

Kepler's *first law* states that *the planets move in elliptical orbits that have the sun at one focus.* An ellipse is one of the figures formed when a cone is cut by a plane (see Figure 4.5). It is also the path traced by a point that moves in such a way that the sum of its distances from two fixed points is constant. The two fixed points are the foci of the ellipse. Thus, in the ellipse in Figure 4.6,

$$P_1 F_1 + P_1 F_2 = P_2 F_1 + P_2 F_2$$

The longest dimension of an ellipse is its *major axis* (a in Figure 4.6). The ratio of the distance between the foci to the major axis is the *eccentricity* of an ellipse; in Figure 4.6, the eccentricity is b/a. Eccentricity is a measure of the shape of an ellipse. As the eccentricity approaches zero, the ellipse approaches a circle in shape. As the eccen-

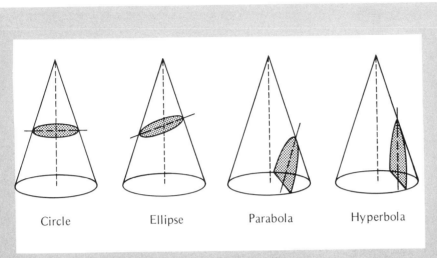

FIGURE
4.5

Circle   Ellipse   Parabola   Hyperbola

Conic sections. Four kinds of curves are formed when a
right circular cone is sectioned by planes passing through it
at various angles: circle, ellipse, parabola, hyperbola.

tricity approaches 1, the shape of the ellipse approaches a straight line.
Figure 4.7 illustrates ellipses of various eccentricities.

Kepler's deduction that planetary orbits are elliptical was a remark-
able achievement. The eccentricity of Mars's orbit, the orbit with which
he worked most, is only 0.093. The other orbits for which he had some
data are even more nearly circular.

Before determining the orbit of Mars, Kepler found the shape of the
earth's orbit. He did this by studying relationships of the earth, sun,
and Mars. He accepted Copernicus's conclusion that Mars orbited the
sun in 687 days. Kepler worked with data gathered by Brahe in a series
of observations made at 687-day intervals; these data are illustrated in
Figure 4.8. In part (a), the sun, earth, and Mars are in a straight line. In
part (b), 687 days later, Mars has completed an orbit of the sun, but the
earth is not where it was in part (a). However, lines of sight to the sun
and Mars at that time intersect at a point on the earth's orbit. A third
point on the earth's orbit can be determined from data on sun and Mars
sightings 687 days later; these data are illustrated in part (c). Using
Brahe's data in this manner, Kepler found the earth's orbit to be nearly
circular, with the sun slightly off center.

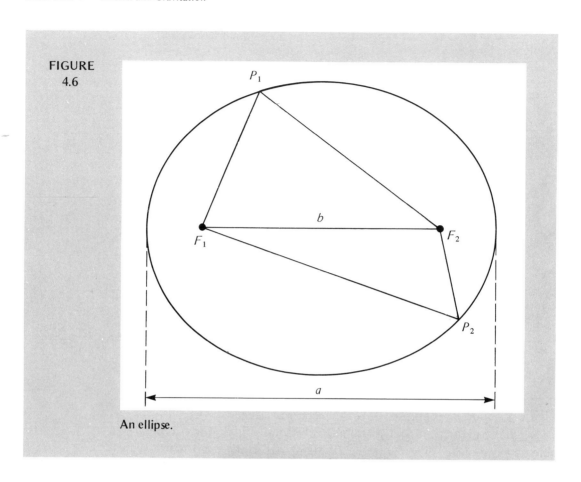

FIGURE
4.6

An ellipse.

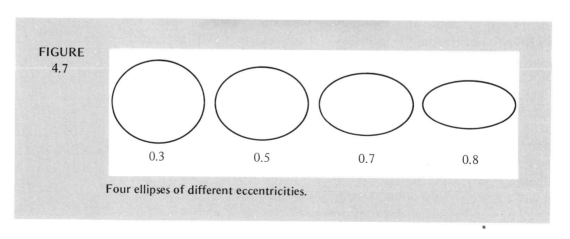

FIGURE
4.7

0.3   0.5   0.7   0.8

Four ellipses of different eccentricities.

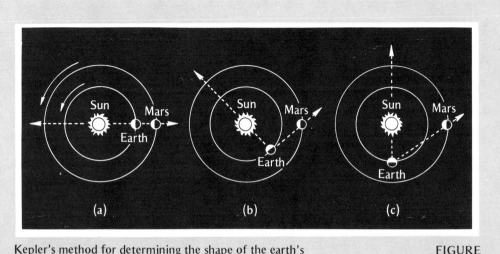

Kepler's method for determining the shape of the earth's orbit.

FIGURE 4.8

It became evident to Kepler that the earth moved at varying speeds in its slightly oval orbit. It moved fastest when it was closest to the sun and slowest when it was farthest from the sun.

Knowing the shape of the earth's orbit and the position of the earth in its orbit throughout the year, Kepler was able to determine the orbit of Mars. Again he used observations of Mars and the sun made at 687-day intervals. The procedure was the reverse of that used in determining positions of the earth in its orbit. This time he knew the earth's position at the beginning and the end of the 687-day interval, and he used lines of sight at the beginning and the end of the interval to determine the position of Mars. This is illustrated in Figure 4.9. Kepler concluded that Mars, like the earth, has a somewhat oval orbit. After several months of additional analysis he determined that the ovals were ellipses.

Kepler's Second Law

Kepler found that the speed of a planet is greatest when the planet is nearest the sun and least when the planet is farthest from the sun. Analysis of this behavior revealed *Kepler's second law: The line from*

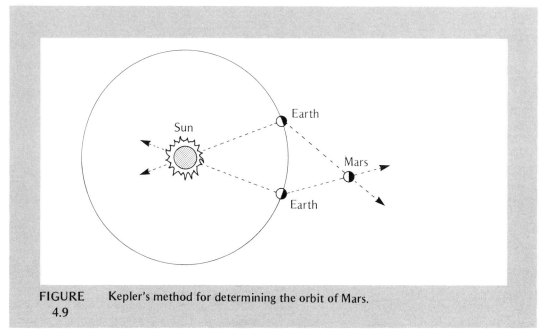

FIGURE 4.9   Kepler's method for determining the orbit of Mars.

the sun to a moving planet sweeps over equal areas in equal intervals of time. This is illustrated in Figure 4.10, in which the eccentricity of the orbit is greatly exaggerated. The shaded portions are equal in area, and the interval during which the planet moves from *A* to *B* is equal to that in which it moves from *C* to *D* and from *E* to *F*.

Kepler's Third Law

The farther a planet is from the sun, the longer it takes to orbit the sun. *Kepler's third law* (the harmonic law) describes this situation: *The squares of the orbital periods of the planets are proportional to the cubes of their mean distances from the sun.* If *P* represents the period of a planet and *D* represents its distance, then

$$P^2 = kD^3 \qquad \text{or} \qquad \frac{P^2}{D^3} = k$$

where *k* is a constant. If time is measured in units of earth years and distance in terms of astronomical units, then

$$\frac{P^2}{D^3} = 1 \qquad \text{or} \qquad P^2 = D^3$$

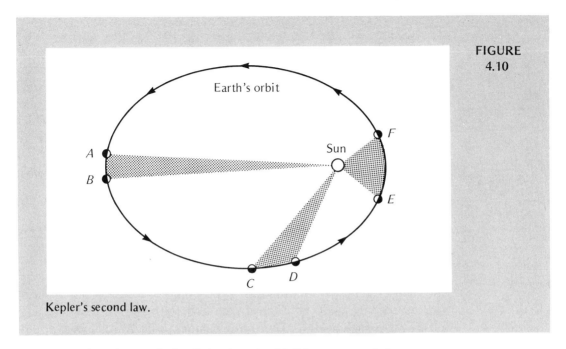

FIGURE
4.10

Kepler's second law.

For example, the period of Jupiter is 11.86 years and its average distance from the sun is 5.2 A.U.; 11.86 squared (140.6) equals 5.2 cubed (140.6).

NEWTON AND UNIVERSAL GRAVITATION

Galileo and Kepler attempted to describe motion rather than explain what forces might cause earthly objects to fall and celestial objects to be pulled into curved paths. Isaac Newton said that all objects pull on one another and that it is this universal force of gravitation that causes an apple to fall and the path of Mars to be curved.

Newton's First Law of Motion

The property of an object that causes it to resist a change in its state of rest or motion is *inertia*. Galileo and Newton reasoned that a ball set rolling at a uniform speed on a smooth, level, and endless surface would continue to roll at that speed forever, if it were not for air resistance and friction. *No* force is being applied to the ball in this case, but it continues in uniform motion. In contrast, an automobile traveling at an unchanging speed is being acted upon by balanced forces—those from the engine pulling it forward and those from friction and air resistance holding it back. If the forces remain balanced, the automobile will not speed up or slow down. Newton's first law describes situations in which either no forces or balanced forces are acting upon an object. It may be stated as follows: *An object under the action of no external force remains at rest or continues to move with uniform speed in a straight line.*

Newton's Second Law

The second law of motion describes the effect of unbalanced forces. Consider a situation in which an experimenter has attached a spring balance to the front of a small cart and is thereby pulling the cart forward (see Figure 4.11). The spring balance indicates that he is pulling with a steady, unbalanced force and that the cart is increasing its speed.

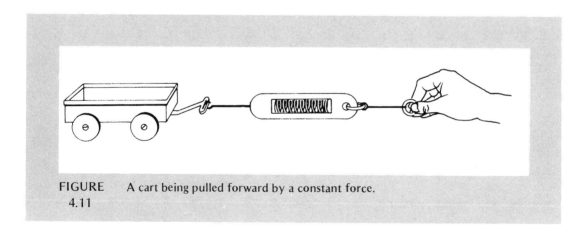

FIGURE   A cart being pulled forward by a constant force.
4.11

Suppose that the cart moves 3 cm during the first second. Subsequent motion of the cart would be as indicated in Figure 4.12. Note that the constant, unbalanced force produces a uniform acceleration.

Next, suppose that the experiment is repeated but with the force doubled; that is, the experimenter pulls steadily, keeping the reading on the spring balance twice as great as in the first case. The motion of the

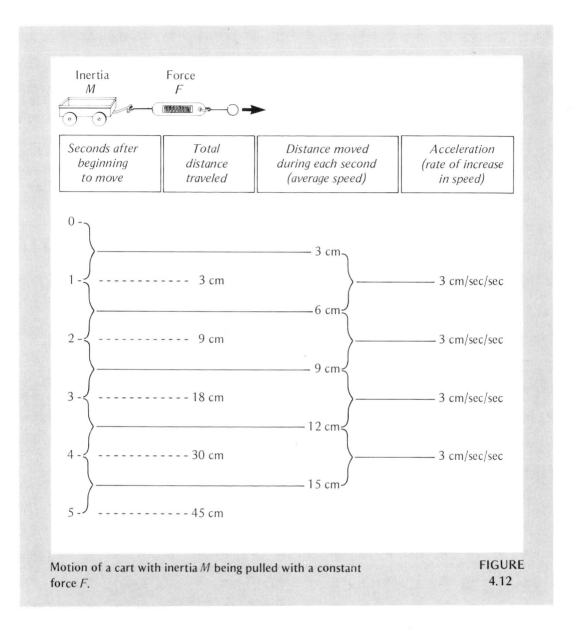

Seconds after beginning to move	Total distance traveled	Distance moved during each second (average speed)	Acceleration (rate of increase in speed)

0 -

--- 3 cm ---

1 - ----------- 3 cm ----------- 3 cm/sec/sec

--- 6 cm ---

2 - ----------- 9 cm ----------- 3 cm/sec/sec

--- 9 cm ---

3 - ----------- 18 cm ----------- 3 cm/sec/sec

--- 12 cm ---

4 - ----------- 30 cm ----------- 3 cm/sec/sec

--- 15 cm ---

5 - ----------- 45 cm

Motion of a cart with inertia M being pulled with a constant force F.

FIGURE
4.12

cart would be as summarized in Figure 4.13. Note that the acceleration is again uniform, but that twice the force has resulted in twice the acceleration.

Suppose that the experiment is repeated once more. This time the original force is used, but the original inertia of the cart is doubled by

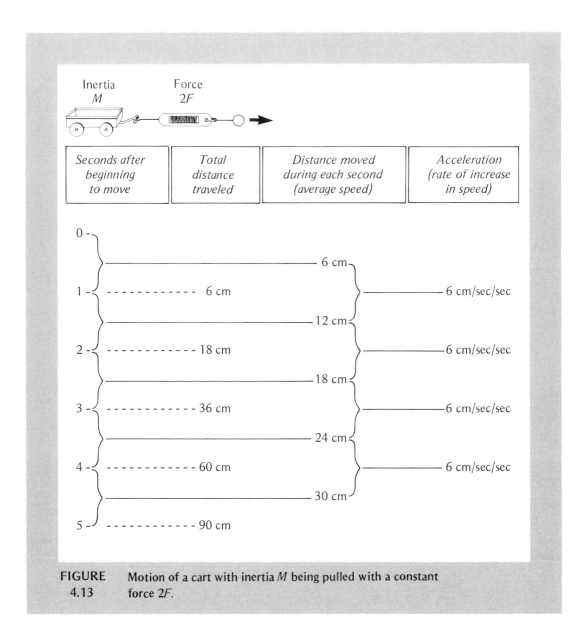

FIGURE   Motion of a cart with inertia *M* being pulled with a constant
4.13   force 2*F*.

placing an appropriate object in it. The motion of the cart is summarized in Figure 4.14. Note that the original force has again produced a uniform acceleration, but with twice the inertia the acceleration is one-half as great as that originally produced by the force.

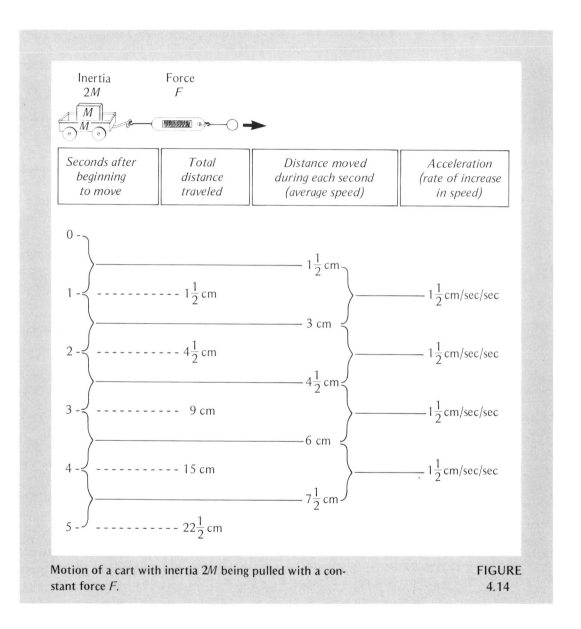

Motion of a cart with inertia 2M being pulled with a constant force F.

FIGURE
4.14

Inertia is measured in terms of *mass.* Mass is not precisely comparable to *weight.* The mass of an object (its inertia or resistance to acceleration) does not vary with its location, but its weight does. A heavy, massive object would weigh much less on the moon than on the earth; however, it would remain as massive on the moon, and therefore as difficult to set in sudden motion as it was on earth.

Newton's *second law of motion* interrelates force, mass, and acceleration as follows: *An unbalanced force acting on a body produces an acceleration of the body directly proportional to the force in the direction of the force and inversely proportional to the mass of the body.* Expressing this as an equation,

$$a = \frac{F}{m} \quad \text{or} \quad F = ma$$

When this equation is used in calculations, units of measurement must be chosen carefully. If the acceleration is to be measured in meters/second^2, mass should be measured in kilograms and force in newtons. One newton (N) is the force required to give a one-kilogram mass an acceleration of one meter per second per second. What acceleration would a force of 10 N give to a 2.5 kg mass?

$$a = \frac{F}{m}$$

$$a = \frac{10 \text{ kg} \cdot \text{m/sec}^2}{2.5 \text{ kg}}$$

$$a = 4 \text{ m/sec}^2$$

Newton's Third Law

Newton's third law concerns the fact that forces are always paired. The *third law of motion* states: *For every action there is an equal and opposite reaction.* In other words, for every force there is an equal and opposite force. When you walk, you push on the earth with your feet and the earth pushes back. Neither of these forces can exist without the other. The second law ($F = ma$) should help make it evident that the earth is exerting a force on you. Since your mass is being accelerated, something must be exerting a force on you. You cannot accelerate yourself—you cannot walk on air or in the void of space. The earth is exerting a force on you and the resulting acceleration can be observed. Similarly, you are accelerating the earth in accordance with the second law, but, since the mass of the earth is so great, the acceleration given to the earth by the push from your feet is imperceptible.

Another example of the third law is the motion of a rocket. This motion is caused by internal gas pushing on the internal surface of the rocket and the surface pushing back on the gas. In Figure 4.15 the throat of the left-hand rocket is closed. The enclosed gas is pushing on the inner surface of the rocket and the surface is pushing back. Since there is no opening through which it can move, the gas cannot be accelerated. The throat of the right-hand rocket is open, and there is no surface there to push back on the gas. The gas is accelerated through the throat in one direction, and the rocket is accelerated in the opposite direction.

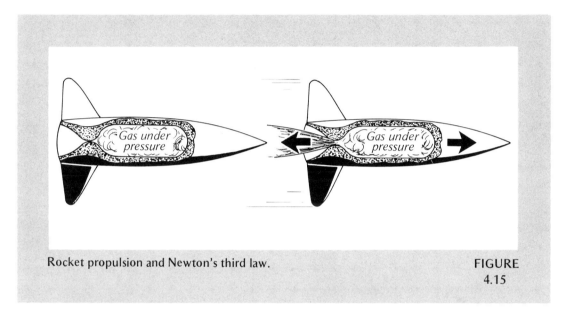

Rocket propulsion and Newton's third law.

FIGURE
4.15

Law of Universal Gravitation

The possibility that there existed an attractive force between masses had been considered prior to Newton. However, it was his thinking on the subject that led to the *law of universal gravitation: An attractive force exists between all bodies in the universe; the magnitude of the force between any two bodies is proportional to the product of their masses and inversely proportional to the square of the distance between their centers.* Expressing this as an equation,

$$F = G\frac{m_1 m_2}{R^2}$$

where G is a constant. Gravitation is a very weak force. At least one of the pair of objects must be very massive before the gravitational attraction between them can amount to much. *Gravitation* is the universal attraction between all bodies everywhere. The term *gravity* is used to refer to gravitational attraction between a celestial body (such as the earth) and objects near its surface.

In order to determine the gravitational constant G, one needs to know the gravitational force between two known masses separated by a known distance. Newton was unable to detect the gravitational attraction between objects in the laboratory, and therefore he could not determine G with accuracy. He could measure the force of attraction between the earth and an object of known mass, because he knew the acceleration the attraction would cause if the object fell. He estimated the mass of the earth; he knew the volume of the earth and assumed that the material composing the earth was, on the average, five times as dense as water—a fairly good estimate. Using the estimated mass of the earth, the known mass of another object on the earth's surface, the distance between the center of the earth and the object, and the attractive force between the earth and the object, he determined G as accurately as he could. About a century after Newton, Henry Cavendish succeeded in detecting the tiny gravitational attraction between some lead balls; his work enabled him to determine the gravitational constant with considerably greater precision than Newton had been able to. The value of G has been refined in subsequent experiments; the value presently accepted for G is

$$G = 6.67 \times 10^{-11} \text{ N} \cdot \frac{\text{m}^2}{\text{kg}^2}$$

A young man who weighs 70 kg is standing 10 m from a young woman who weighs 50 kg. What is the attraction between them?

$$F = G\frac{m_1 m_2}{R^2}$$

$$F = 6.67 \times 10^{-11} \text{ N} \cdot \frac{\text{m}^2}{\text{kg}^2} \times \frac{70 \text{ kg} \cdot 50 \text{ kg}}{(10 \text{ m})^2}$$

$$F = 6.67 \times 10^{-11} \text{ N} \cdot \frac{\cancel{\text{m}^2}}{\cancel{\text{kg}^2}} \times \frac{3,500 \cancel{\text{kg}^2}}{100 \cancel{\text{m}^2}}$$

$$F = 2.3 \times 10^{-9} \text{ N}$$

A force of 2.3×10^{-9} newtons is less than 1/1000 the weight of a postage stamp!

Centripetal Acceleration

Acceleration has been defined as a change in either the speed or direction of motion; a force is required to accelerate a body. Since the planets orbit the sun rather than moving off into space along straight lines, it is evident that a force (gravitation) is accelerating them. The earth and the sun are pulling on each other with equal and opposite forces along a line joining their centers. Since the sun is vastly more massive than the earth, the earth is accelerated more than the sun and is pulled into orbit around the sun. The earth is accelerated *toward* the sun, and the effect of the acceleration is to curve the path the earth takes. This continual acceleration toward the sun prevents the earth's

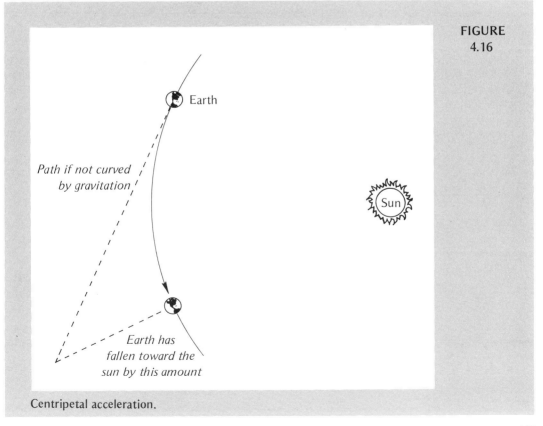

FIGURE
4.16

Earth

Path if not curved
by gravitation

Sun

Earth has
fallen toward the
sun by this amount

Centripetal acceleration.

inertia from moving the earth along a straight line. Conversely, the earth's inertia prevents it from falling into the sun. (See Figure 4.16.) Acceleration toward the center of curving motion is called *centripetal acceleration.*

Gravitational Force and Distance

Newton was able to compare the known acceleration of objects falling near the earth's surface (about 10 m/sec^2) with the centripetal acceleration of the moon, because he knew the dimensions of the moon's orbit and knew that it takes the moon about $27\frac{1}{3}$ days to orbit the earth. He found that the moon's acceleration toward the earth is about 10 m/min^2. That is, the moon is pulled away from a straight path and curved toward the earth at this rate. This does not mean that the moon is about 10 m closer to the earth at the end of each minute than it would be if it had traveled in a straight line. The actual distance is half that. An acceleration of 10 m/min^2 indicates that a velocity of 10 m/min toward the earth develops during each one-minute interval. The average velocity during the minute is 5 m/min. The ratio of these two accelerations (10 m/min^2 to 10 m/sec^2) is 1/3600. Newton knew that the distance between the center of the earth and the center of the moon is about 60 earth radii. This indicated to Newton that the gravitational forces between objects diminish as the square of the distance between the centers of the objects (1/3600 the force at 60 times the distance).

Gravitational Force and Mass

The third law tells us that all forces are paired; therefore, gravitational forces are paired. A pebble dropped to the ground pulls on the earth as hard as the earth pulls on the pebble. The acceleration of the pebble is obvious, whereas that of the massive earth cannot be detected. Gravitational force exists only when two or more masses pull on one another. The gravitational attraction between two objects is proportional to the product of their masses. If the mass of one object in a pair is increased by five times, the gravitational attraction between them is increased by five times. However, if the masses of both objects in a pair are increased by five times, their gravitational attraction is increased by twenty-five times.

Kepler's Ellipses as Accounted for by Newton

Newton was able to demonstrate mathematically that his laws of motion and gravitation account for the elliptical orbits of the solar system—that they are as applicable to elliptical orbital motions as to circular orbital motions. In order to do this, he had to invent calculus.

Galileo and Kepler described the motions of falling and orbiting bodies, but not the forces causing them. Newton described the forces that cause the motions. But did Newton complete the explanation? Why is there gravitational attraction between all bodies?

Complex Gravitational Interactions

Gravitational forces are negligible except between pairs of objects that include at least one celestial body. And even these forces diminish rapidly as the distance between the bodies increases. Still, the gravitational interactions among solar system bodies are complex, since there are many massive bodies close enough together to tug on one another. *Perturbations* are the minor deviations in the orbital motion of a body caused by these complex gravitational relationships.

The foregoing discussion of orbital motion suggests that one object orbits around another. This is an oversimplification. Actually, the two objects revolve around the center of mass. The center of mass is on the line joining the centers of the objects. It is located at the point where the product of the mass times the distance to the center of the object at one end of the line is equal to the product of the mass times the distance to the center of the object at the other end. In solar system orbits, the effect is essentially that of one object orbiting the other, because one member of the pair is much more massive than the other. The earth and moon revolve around a center of mass that is within the earth. The sun and planets revolve around centers of mass within the sun.

"Artificial" Orbits

The concept of orbiting a man-made object around the earth dates from Newton, not Sputnik. In his *Principia*, Newton said that one could imagine firing a cannonball from a mountain a thousand miles high with such a high velocity that it would be over the curved rim of the earth too quickly to fall back to earth. It would then be in orbit around the earth (see Figure 4.17).

FIGURE
4.17

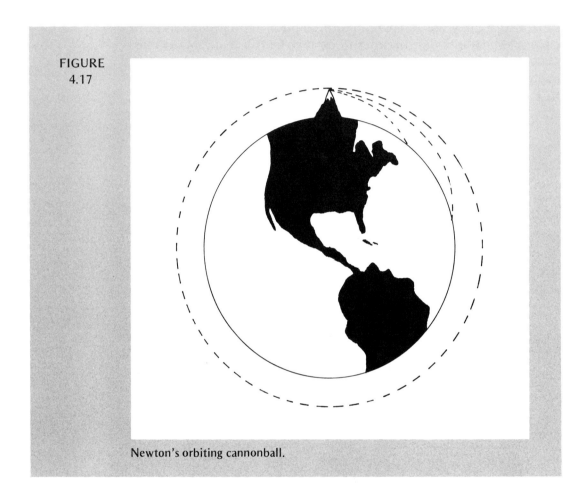

Newton's orbiting cannonball.

In effect, we do this each time we place an artificial satellite in orbit. A rocket lifts the satellite upward, turns, and hurls the satellite over the rim of the earth. If the satellite is relatively near the earth (approximately 160 kilometers away) the speed necessary to put it into orbit around the earth is about 29,000 km/hr (about 18,000 mph). If an object can be hurled away from the earth at about 40,000 km/hr (25,000 mph), it will break away from the earth's pull. It will then go into orbit around the sun, unless it approaches close enough to another solar system object to be "captured" by its gravitation.

EINSTEIN AND GRAVITATION

Despite the sweeping power of Newton's laws to describe and predict, some shortcomings had become evident by the beginning of this century. For instance, astronomers and physicists were perplexed by the fact that the speed of light appeared to be the same when measured in all directions relative to the earth's motion. Also, Mercury's orbital motion seemed to be perturbed slightly more than could be expected from Newtonian gravitational mechanics. Albert Einstein published his *special theory of relativity* in 1905; by 1916 he had extended this work to formulate his *general theory of relativity*. General relativity theory suggested new laws of motion and gravitation which predict motions at speeds approaching that of light more accurately than Newtonian laws.

CONCLUSION

Newton's work enables us to explain the falling of an apple, the trajectory of a projectile, tides, the mechanics of the solar system, motions of objects in distant space, and many other phenomena. The complex calculations necessary to plan each space trip are based on fundamental principles revealed by Galileo, Kepler, and Newton. Newton had considerable influence in such areas as philosophy, literature, and political theory, because many people hoped that reasoning such as his could reveal fundamental principles that would guide and explain human affairs.

FOR REVIEW, DISCUSSION, OR FURTHER STUDY

1. Across what distance will the rolling ball in Figure 4.3 travel during the twelfth second after release?

2. Determine whether or not terms are used correctly in the following statement: The velocity of an automobile is 60 kilometers per hour.

3. What is the comparative gravitational attraction between two objects before and after the distance between them is quadrupled?

4. Is the orbital speed of the earth in January greater or less than its speed in June?

5. If the mean distance to the moon were less, would the moon's mean orbital speed be more or less than it is now?

6. The gravitational attraction between two objects separated by a given distance would be greater if the mass of either object were greater. If the moon were more massive than it is, would it necessarily orbit the earth more closely than it does now?

7. There is a common misconception that an outward ("centrifugal") force balances the attractive force between a pair of orbiting objects such as the earth and the moon. Explain why this is erroneous.

8. Why are space vehicles ordinarily launched in an easterly direction?

9. Some artificial earth satellites are placed in orbits such that they always remain above the same spot on the earth's surface. In which direction are such satellites moving, and what is the shape of their orbits?

10. An object weighing 1 kg dropped from the roof of a building will fall about 4.88 m in the first second. If the moon were stopped in its orbital motion and then allowed to fall, would it fall more or less than 4.88 m in the first second?

SUGGESTED ACTIVITIES

Acceleration of Freely Falling Bodies

Repeat an experiment of Galileo: Drop balls of unequal weight simultaneously from a second-story window. Observe that they reach the ground at the same time.

Forward Motion and the Acceleration of Freely Falling Bodies

Place two balls (such as marbles or ball bearings) at the edge of a table. Simultaneously flick both of them from the table, one with one index finger and the other with the other index finger. Impart much greater forward velocity to one ball than to the other. A few minutes of practice may be required to develop the necessary dexterity, but the point of the demonstration should soon become evident: If a projectile is fired horizontally and an object is simultaneously dropped vertically from the same starting point, they will reach the ground simultaneously.

Some Orbital Mechanics

Obtain about three feet of strong twine. Securely tie an object weighing about 2 or 3 oz to one end of the twine. With one hand grasp the twine about $1\frac{1}{2}$ ft from the object; hold the end of the twine with the other hand. Hold the dangling object in front of you at arm's length and waist height. Swing the object in a circular path. While the object is "orbiting," draw it closer to your hand by pulling the twine through the fingers that are supporting the object. This is easier to do if you pass the twine through a metal washer which you hold between your thumb and index finger, but a washer is not necessary. A few minutes of practice may be necessary to develop the required dexterity, but it should soon become evident that the object orbits faster when a stronger attractive force draws it closer to your hand.

SUGGESTED READINGS

Canby, Thomas Y., "Skylab: Outpost on the Frontier of Space," *National Geographic*, Vol. 146, No. 4 (October 1974), 441.

Drake, Stillman, "Galileo's Discovery of the Law of Free Fall," *Scientific American*, Vol. 228, No. 5 (May 1973), 84.

Drake, Stillman, "Galileo's Discovery of the Parabolic Trajectory," *Scientific American*, Vol. 232, No. 3 (March 1975), 102.

Driscoll, Everly, "A Planet for the Sun's Twin," *Science News*, Vol. 103 (10 February 1973), 91.

This brief article reports perturbations in the motion of Epsilon Eridani and suggests that a planet causes them.

Einstein, Albert, "On the Generalized Theory of Gravitation," *Scientific American*, Vol. 182, No. 4 (April 1950), 13.

Gamow, George, "Gravity," *Scientific American*, Vol. 204, No. 3 (March 1961), 94.

Garriott, Owen K., "Skylab Report: Man's Role in Space Research," *Science*, Vol. 186 (18 October 1974), 219.

Logan, Jonathan L., "Gravity Waves—A Progress Report," *Physics Today*, Vol. 26, No. 3 (March 1973), 44.

Robinson, Arthur L., "Relativity: Experiments Increase Confidence in Einstein," *Science*, Vol. 188 (13 June 1975), 1099.

Thomsen, Dietrick E., "Which Theory of Gravity Fits?" *Science News*, Vol. 102 (29 July 1972), 76.

Van Flandern, Thomas C., "Is Gravity Getting Weaker?" *Scientific American*, Vol. 234, No. 2 (February 1976), 44.

Will, Clifford M., "Gravitation Theory," *Scientific American*, Vol. 231, No. 5 (November 1974), 24.

Wilson, Curtis, "How Did Kepler Discover His First Two Laws?" *Scientific American*, Vol. 226, No. 3 (March 1972), 93.

FIGURE
5.1
Objective prism spectrogram of dozens of stars.
(Photograph courtesy of Department of Astronomy,
The University of Michigan.)

RADIATION:
The Stream of Energy and Information from Space

5

A few hundred pounds of moon rocks and a few tons of meteorites are the only extra-terrestrial materials available for study. Yet astronomers speak confidently of the physical and chemical properties of bodies that appear only as points of light in telescopes. In many cases the points of light cannot be seen; they appear only on photographic plates exposed for hours in the largest telescopes. In some cases astronomers describe objects that can be detected only with gigantic radio telescopes. The radiant energy coming to the earth through the void of space is the source of nearly all of our knowledge of the universe. How can we learn so much from this energy?

WHAT IS LIGHT?

What is light? This question puzzled the world for centuries. Newton did significant work in optics, and he believed light to be a stream of particles or corpuscles. Some other scientists of the time believed it to be waves of some sort. Controversy over these two views of the nature of light continued for decades. Some attempted to explain the behavior of light by visualizing a wave model; others tried to use a particle model. By the end of the last century, the wave model had been generally accepted; the behavior of light had been explained in terms of wave properties. Early in this century Albert Einstein, building upon work done by Max Planck, explained the photoelectric effect; this effect is the emission of electric charges from metal that has been struck by light. His explanation required that light waves behave as though they were particles. These particles were termed *quanta* or *photons.* Light is currently viewed as having a dual nature—waves that sometimes act like particles and vice versa. Most find it difficult to visualize a physical model of this wave-particle duality. Physicists deal with a mathematical model of the duality in the field known as *quantum mechanics.*

ELECTROMAGNETIC ENERGY

All of us have seen the effects of magnetic and electric fields. Magnets can move objects without touching them. When you bring a comb charged with static electricity close to your head, it will cause your hair to rise.

In 1800 Alessandro Volta discovered how to produce an electric current from a cell or battery of cells. A cell consists of two materials, such as copper and zinc, placed close together but separated by a salt solution or a similar liquid. In 1820 Hans Christian Oersted discovered that an electric current, a moving charge, is surrounded by a magnetic field. So electricity can produce magnetism. Again, examples of this phenomenon are commonplace. Electromagnets are present in electric motors, doorbells, and other familiar devices.

In 1831, Michael Faraday made the profoundly important discovery that the converse of this situation is true; he discovered that an electric current can be produced by moving a wire in a magnetic field. This phenomenon, called *electromagnetic induction,* is utilized in the generators that produce the electric current used in our homes and businesses.

James Clerk Maxwell, a brilliant theoretician, suggested in the 1860s that a changing electric field (not just an electric current) produces a magnetic field. He utilized the concepts of electric and magnetic fields in developing the theory that an accelerated electric charge (one moving with a changing velocity) would produce both a changing or disturbed electric field and a changing or disturbed magnetic field. He said that these changing fields would sustain one another; the changing magnetic field would continue to induce a changing electric field, and the changing electric field would induce a changing magnetic field. These mutually self-sustaining disturbances in fields would move through space as waves of *electromagnetic energy*.

In the late 1880s, Heinrich Hertz was able to generate and detect invisible electromagnetic waves and to demonstrate that they had many of the properties of light waves. The experiments of Hertz confirmed predictions based on Maxwell's theory. Thus we have been led to the conclusion that light is made up of waves of electromagnetic energy—as are radio waves, infrared waves, ultraviolet waves, x-rays, and gamma rays.

The Electromagnetic Spectrum

Figure 5.2 may help you to visualize the characteristics of waves. The *wavelength* (symbolized by the Greek letter lambda) is the distance from crest to crest or trough to trough. The *amplitude* is the height of a crest or the depth of a trough. The *frequency* is the number of crests or troughs that pass a given point in a unit of time. The velocity of a wave equals the frequency times the wavelength. For example, if 3 waves of water, each 2 cm long, pass a given point in one second, the velocity of the wave is 6 cm per second. These relationships among wavelength, frequency, and velocity are expressed by the equation $\lambda = v/f$.

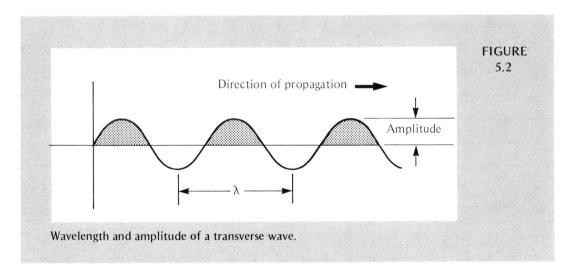

FIGURE 5.2

Wavelength and amplitude of a transverse wave.

FIGURE
5.3

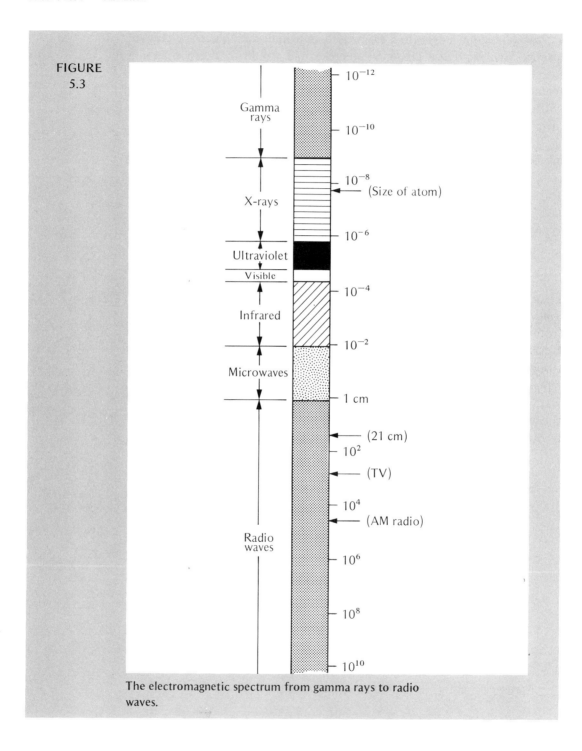

The electromagnetic spectrum from gamma rays to radio waves.

The speed of electromagnetic energy in a vacuum is constant. This speed is 2.9979 $\times$ 10^8 m/sec, which is about 186,000 miles per second or 300,000 kilometers per second. Chapter 2 described how Römer made the first good estimate of the speed of light nearly 300 years ago. Since the speed of electromagnetic energy is constant, only the frequency and wavelength can vary. And they do vary over an immensely wide range. Obviously, the longer the wavelength, the lower the frequency; and the shorter the wavelength, the higher the frequency. The longest known electromagnetic waves are a few kilometers in length, and the shortest have lengths considerably less than the diameters of atoms.

Differing wavelengths and frequencies result in strikingly different effects of the radiation, even though all electromagnetic radiation is intrinsically similar. Radio waves are the longest; and gamma rays, which are emitted in nuclear reactions, are the shortest. The continuous range of wavelengths and frequencies is the *electromagnetic spectrum.* (See Figure 5.3.) Visible light is a very small portion of the electromagnetic spectrum. Light waves vary in length from about 0.0004 cm to about 0.0007 cm, and they have frequencies from about 4 $\times$ 10^{14} to about 8 $\times$ 10^{14} cycles per second. Essentially all of the wavelengths in the electromagnetic spectrum are emitted by various celestial bodies, but the earth's atmosphere is opaque to most of them. The *optical window* permits light to reach the earth's surface, and a wider *radio window* is transparent to those radio waves with lengths from about 1 cm to 15 m.

Color and Wavelength

The color of light is determined by its wavelength. The longest electromagnetic waves that the eye detects as light are red in color. The colors associated with progressively shorter wavelengths range from orange through yellow, green, and blue to violet. A unit for measuring the length of light waves is the angstrom (A); one angstrom is equal to 10^{-8} cm. Light waves range in length from about 4,000 A (violet) to about 7,000 A (red).

White light includes waves of many different lengths—light of all of the colors from red to violet. This is readily demonstrated by passing white light through a prism, as in Figure 5.4. The light emerges from the prism spread out into a spectrum of the six colors of the rainbow (in a rainbow drops of water act like prisms).

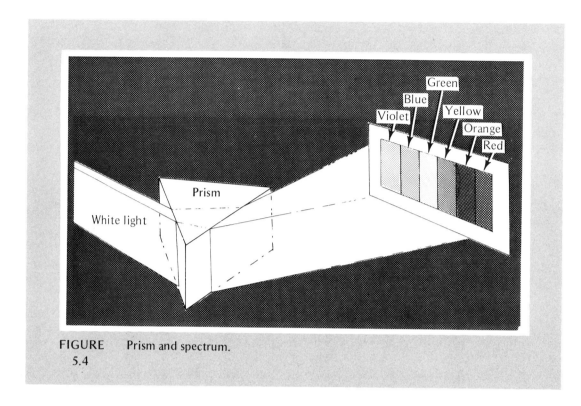

FIGURE   Prism and spectrum.
5.4

REFRACTION AND REFLECTION

A prism disperses the colors in white light by slowing down the light passing through it. When the speed of light is slowed by the glass, the different colors of light are slowed unequally—violet is slowed the most. When light passes obliquely from a transparent medium of one density into a medium of different density, its speed and direction are changed; the path of the light is bent or *refracted*. Refraction does not occur if the light enters a medium in a direction perpendicular to the medium's surface. The amount of refraction that will take place when light enters a particular medium or material is indicated by the *index of refraction* of the material. The index of refraction of a given material differs for different wavelengths (colors) of light. The indices of refraction vary for different kinds of glass. The index of refraction of air also varies with the temperature, pressure, and water vapor content of the air.

Parts (a) and (b) of Figure 5.5 illustrate the effect of a prism on monochromatic light and on white light. If the index of refraction of

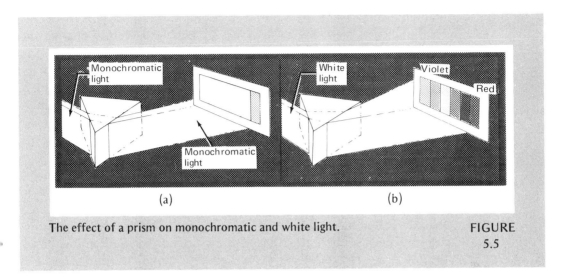

The effect of a prism on monochromatic and white light.

FIGURE
5.5

the glass did not differ for different wavelengths, the prism would only bend white light and not disperse its colors.

Reflection is another means by which the direction of the path of light is changed. The surfaces of some materials reflect most of the light striking them. Reflected light rays travel in a direction that is related specifically to the direction of the incident rays. The term *ray*, as used here, means a very narrow beam of light; it is a convenient concept for considering the paths of reflected and refracted light.

RADIATION AND ATOMIC STRUCTURE: THE BOHR ATOM

Maxwell explained that electromagnetic radiation is emitted when an electric charge is accelerated. Does this explanation account for starlight? Where are the accelerating charges that generate the light, radio, and other radiation with which the astronomer works? In order to answer these questions, we must consider some aspects of atomic structure.

The tiniest particle of each of the 92 natural elements is an atom. Atoms are made of even more fundamental particles—principally protons, electrons, and neutrons. The numbers of protons and electrons are equal and range from one of each in a hydrogen atom to 92 of each in a uranium atom. There are varieties or isotopes of atoms, which differ in the number of neutrons. Protons have a positive electrical charge, and electrons have an equal but negative charge. The protons and neutrons, which are close together in the atomic nucleus, provide nearly all the mass of the atom. The electrons are outside of the nucleus and in exceedingly swift motion about it. All atomic dimensions are extremely small, but the relative distances from the nuclei to the electrons are analogous to those from the sun to the planets. Atoms are mostly empty space.

In 1911, Ernest Rutherford developed a model in which the atom resembled the solar system. He pictured electrons orbiting the nucleus as planets orbit the sun with the attraction between the positive electric charge of the nucleus and the negative charge of the electrons functioning like gravitational attraction in the solar system. Coulomb's force— the force of attraction that exists between unlike electric charges and the force of repulsion between like charges—was known of prior to Rutherford's day; Rutherford incorporated it into his model. In a relationship similar to that of gravitational forces, the Coulomb force varies directly as the product of the charges and inversely as the square of the distance between them. Relatively speaking, Coulomb forces are much stronger than gravitational forces. Now the electrons are believed to surround the nucleus in a vibrating cloud of electric charge, so Rutherford's solar system analogy is considered to be an oversimplification.

Does the concept of moving electrons begin to suggest the source of light? If electrons are moving electric charges, doesn't it follow that acceleration of their motion would give rise to electromagnetic radiation in accordance with Maxwell's explanation? But isn't there something contradictory in this? In the last chapter acceleration was described as a change in the direction of motion. If electrons are whirling and vibrating around the nuclei of atoms, their direction of motion is changing continually and they are being accelerated. Shouldn't all atoms then radiate continually until they have given up their energy to electromagnetic waves and until their electrons have been drawn into collapsed atoms? Why isn't this paper emitting a glow? Isn't there a conflict here between Newtonian mechanics and electromagnetic theory? This conflict, which became evident when Rutherford suggested his atomic model, was resolved by Niels Bohr.

Max Planck had stated in 1900 that an atom does not radiate nor absorb energy in a continuous, smooth manner; rather, energy is absorbed or released in pulses of discrete amounts, which are termed *quanta.* These quanta are also the *photons* referred to by Einstein. A single quantum of energy is always minute. Planck stated that the amount of energy in a quantum is proportional to the frequency of the radiation in accordance with the following equation:

$$E = hf$$

E is the energy of the quantum or photon, f is the frequency, and h is Planck's constant. Thus the quanta associated with a particular frequency (wavelength) are always equal.

The equation $E = hf$ suggests the dual wave-particle nature of light. Pulses or quanta of energy imply a particulate or corpuscular model, yet frequency is related to a wave model, as in the relationship

$$v = f \lambda$$

In 1913 Bohr applied Planck's concept of the discontinuous radiation of quanta to the Rutherford atom to form a new atomic model. The Bohr atom provides a very satisfying explanation of the emission and absorption of energy by hydrogen and other simple atoms. Bohr said that there are only certain specific orbits in which electrons can circle the nucleus of an atom; they can be only at certain specific distances and possess specific amounts of energy. As long as the electrons remain in a given arrangement, the atom neither radiates nor absorbs energy. Each orbit is called an *energy level.* Successively more distant orbits are successively higher energy levels; therefore, electrons in more distant orbits possess more energy than those in closer orbits. The electrons must absorb energy to move to higher levels. When electrons drop from higher to lower energy levels, they lose energy, which creates electromagnetic waves. The radiation and absorption occur suddenly, not continuously. The amount of energy abruptly radiated or absorbed when an electron moves to an adjacent energy level is always equal to *hf;* that is, it is a quantum of energy of the particular frequency. So, in the Bohr atom, electromagnetic waves are not generated by the acceleration of electrons—at least not as we previously thought of acceleration. Rather, it is the sudden drop in energy levels of the electrons that produces light and other radiation.

The quantum mechanics of orbiting electrons is different from the Newtonian mechanics of orbiting planets. Celestial mechanics does not say that a particular planet could orbit the sun in only a few different orbits at very specific distances. Depending upon its speed of revolution, a planet could orbit the sun anywhere within the sun's effective gravitational field.

The Bohr model accounts nicely for much of the observed behavior of the simpler atoms as they radiate and absorb energy. It also accounts for many of their chemical properties. The Bohr model cannot account satisfactorily for many observations involving more complex atoms; still, the primitive quantum mechanics of the Bohr atom are helpful in considering optical spectra.

SPECTROSCOPY

The pretty rainbow-colored band of sunlight emerging from a prism reveals an enormous amount about the sun. Careful inspection of the solar spectrum indicated the presence of helium in the sun many years before that element was found on earth. Spectra from points of light have provided much of our information about the universe. How can the study of spectra be so useful to astronomers?

Spectroscopes

The instrument for producing and viewing a spectrum is called a *spectroscope.* The instrument for photographing a spectrum is a *spectrograph.* The optical parts of a spectrograph are illustrated in Figure 5.6.

A narrow slit at one end of the spectrograph in Figure 5.6 is placed in the telescope at the point where the image under study is focused. Only light from this single object will pass through the slit into the spectrograph. The collimating lens renders the rays parallel before they pass into the prism. The objective lens of the spectrograph focuses the spectrum formed by the prism onto the photographic plate. The spectrograph is often designed so that a comparison spectrum can be produced within it and photographed beside that of the star; the comparison spectrum might be that of incandescent iron vapor in an electric arc.

Another type of spectrograph, an *objective prism spectrograph*, has a large narrow prism in front of the telescope so that light passes through the prism before entering the telescope. This arrangement eliminates the slit and collimating lens (Figure 5.7). The photograph in this case, the *spectrogram*, will be of the spectra of all stars that were in

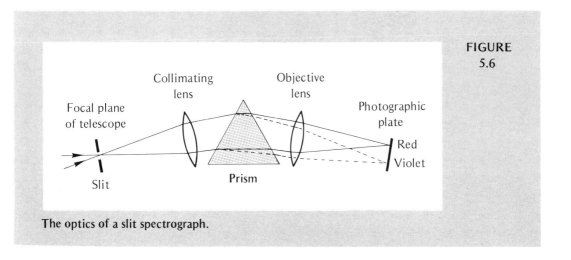

FIGURE
5.6

The optics of a slit spectrograph.

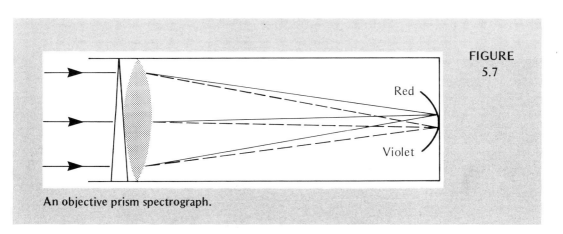

FIGURE
5.7

An objective prism spectrograph.

the telescope's field of view, not of the spectrum of a single star. Spectrograms from objective-prism spectrographs are less detailed than those from a normal spectrograph, but they can provide information about hundreds of stars in a single photo. (See Figure 5.1.)

A diffraction grating may be substituted for a prism in a spectrograph. Diffraction gratings are surfaces on which many very fine grooves have been ruled—perhaps 8,000 or more per centimeter. Some gratings are transparent and others are reflective. In a *transmission grating* the light passes through the transparent material between the grooves; in a *reflection grating* the light is reflected from the surface

FIGURE
5.8

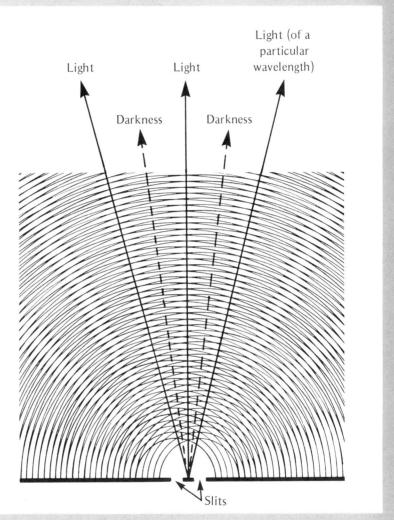

Diffraction. Waves spreading out from the two slits meet destructively in places, leaving darkness. All of the waves illustrated are of the same length (same color); the radii of the arcs drawn around each slit increase by the same amount. If waves of various lengths (colors) were spreading out, the directions from the slits in which they would meet destructively and constructively would vary according to wavelengths; the light would be separated into a spectrum.

between the grooves. The light that is transmitted or reflected is *diffracted*. Diffraction is a wave phenomenon. When waves pass the edges of obstacles, diffraction causes them to spread into the regions behind the obstacles; in effect, waves go around corners because of diffraction. The grooves act as obstacles to the light. In the smooth areas the light is reflected or transmitted. When it is diffracted by the edges of the grooves, the light spreads out from each of the smooth places. Waves spreading out from these smooth places interfere with one another in a definite pattern. Where troughs and crests meet there is *destructive interference;* in the case of light waves, destructive interference leaves darkness. Where crests come together with other crests and troughs with troughs, waves reinforce one another through *constructive interference;* in the case of light waves, this results in increased brightness. The interference pattern depends upon the lengths of the waves. A spectrum is formed because there is a different interference pattern for each color—brightness and darkness occur in a different place for each color. (See Figure 5.8.)

Kinds of Spectra

If the solar spectrum is examined with a slit spectroscope, it seems to be crossed by dozens of narrow, dark lines. The spectrum from the incandescent filament of a light bulb does not have such dark lines; it is a continuous rainbow-colored band. The spectrum from a glowing neon sign is very different from either of these; it consists only of several narrow, bright lines of different colors against a dark background. These are examples of the three kinds of spectra. In 1859, Gustav Kirchhoff identified the circumstances under which each of the three kinds of spectra is produced:

1. A *continuous spectrum* results from light emitted by an incandescent solid, liquid, or gas under high pressure.
2. A bright line or *emission spectrum* is produced by light from an incandescent, low-pressure gas.
3. A dark line or *absorption spectrum* results when light from a source with a continuous spectrum passes through intervening gas that is at a lower temperature and pressure.

These situations are illustrated in Figure 5.9.

FIGURE
5.9

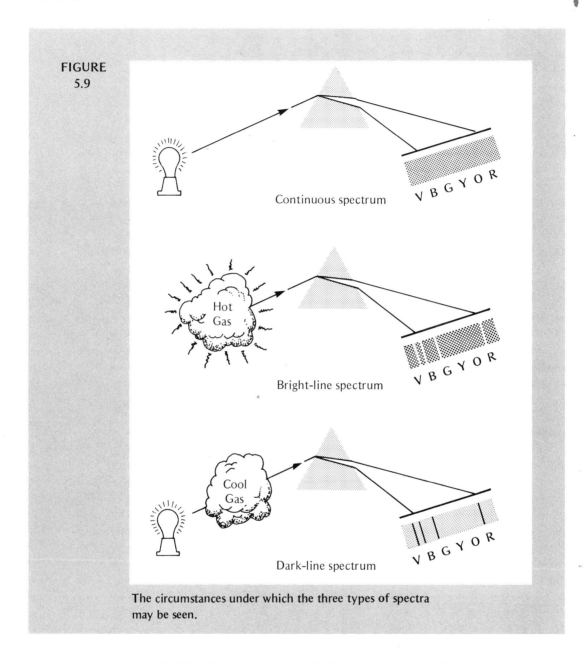

Continuous spectrum   V B G Y O R

Bright-line spectrum   V B G Y O R

Dark-line spectrum   V B G Y O R

The circumstances under which the three types of spectra
may be seen.

Kirchhoff made one particularly important discovery. He noted
that dark lines introduced in a continuous spectrum by passage of the
light through intervening gas are located in the same relative positions

as the bright lines emitted by the same kind of intervening gas when it incandesces under low pressure. That is, the lines—bright or dark—produced by a low-pressure gas are always at the same wavelengths. Since spectrograms can be calibrated accurately, the location of lines along the 4,000–7,000 A band of wavelengths indicates the material that produced the lines (Figure 5.10).

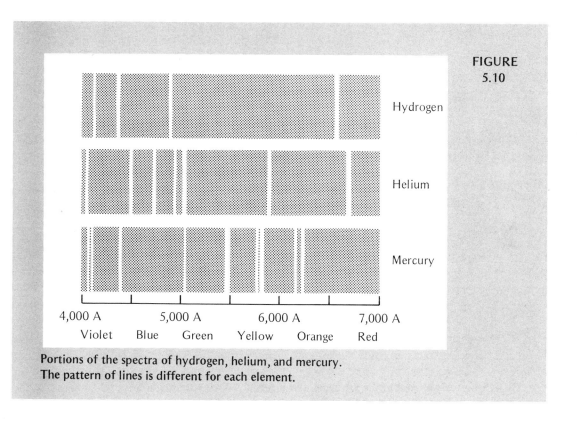

FIGURE 5.10

Portions of the spectra of hydrogen, helium, and mercury. The pattern of lines is different for each element.

The light of the sun is coming from incandescent, high-pressure gas. The dark lines in its spectrum result from the fact that the light passes through less dense outer regions as it leaves the sun. Stars typically produce absorption spectra. Planetary nebulae and some bright diffuse nebulae produce emission spectra. Those bright nebulae that are merely reflecting light from neighboring stars give the dark-line spectra of those stars. Since light from cosmic sources has passed through the earth's atmosphere before arriving at the observatory, all astronomical spectrograms made at the earth's surface include absorption lines caused by the earth's atmosphere. Astronomers must take atmospheric effects into account in interpreting spectra.

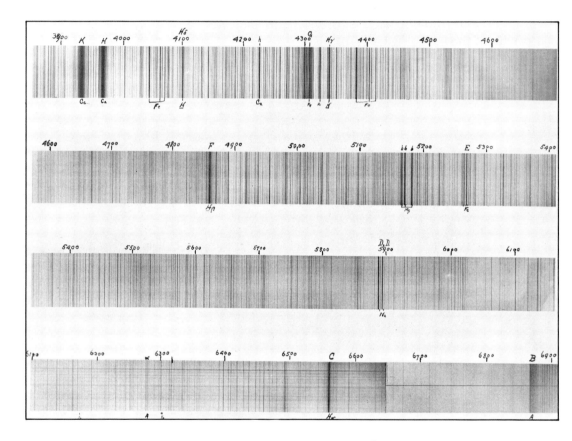

FIGURE 5.11   Spectrum of the sun from 3,900 A to 6,900 A. (Photograph courtesy of the Hale Observatories.)

Spectral Lines and the Bohr Atom

The origin of spectral lines is most easily understood in terms of Bohr's model. Electrons that have been raised to higher energy levels by absorption of energy spontaneously drop to lower levels upon the emission of light. The wavelengths of light emitted in transitions between specific energy levels in specific kinds of atoms are always the same; thus emission lines resulting from particular transitions always appear in the same places in the spectral band. Figure 5.12 diagrams some of the possible transitions to lower energy levels of the electron in a hydrogen atom.

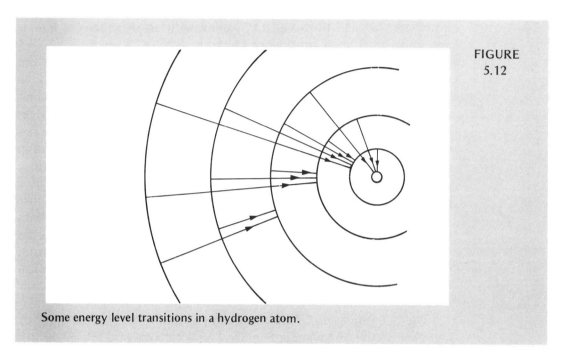

FIGURE
5.12

Some energy level transitions in a hydrogen atom.

As electrons are raised to higher energy levels, atoms absorb photons or quanta of the same energy content that they will later radiate. It is absorption of energy that introduces dark lines. When light from a source with a continuous spectrum encounters low-pressure cool gas, some of the light does not pass on through the gas. Some photons are absorbed by the intervening atoms. The missing wavelengths leave dark absorption lines. The lines are not totally dark for there is still some light at these wavelengths, but they appear dark by contrast. Electrons in the excited intervening atoms return quickly to their original lower energy levels, reradiating the photons they have absorbed. Before absorption these photons were all coming from the direction of the source of the continuous spectrum. They are reradiated in all directions by the intervening atoms; with less light of particular wavelengths moving in the original direction, relative darkness—absorption lines—results at particular places in the spectral band.

The Bohr atomic model seems to account readily for emission and absorption lines. But how is a continuous spectrum produced? At low pressures, gases cause bright and dark spectral lines at positions characteristic of the kinds of atoms in the gases. When these same gases incandesce at high pressure, they—along with all incandescent liquids and solids—produce a continuous spectrum. Why?

Atoms are in constant motion. The pressure of a gas results from collisions of the atoms with the surfaces enclosing the gas. If a low-pressure gas is compressed, its atoms are crowded closer together. Under these circumstances, atoms are influenced by the electric charges in neighboring atoms. Our description of radiation and absorption has been in terms of an isolated atom. The proximity of electric charges in other atoms can influence the amount of energy associated with a particular transition. Photons of differing wavelengths are emitted by a particular kind of energy level transition in the crowded atoms of dense materials. As the pressure of a radiating gas increases, its spectral lines begin to smear—they no longer remain crisply defined at specific wavelengths. There are photons of all wavelengths in the light from dense materials. Since there are then lines at every point of the spectral band, they overlap to form the rainbow-colored band of the continuous spectrum.

The determination of chemical composition from spectral lines is a complex matter. Atoms containing several electrons permit many different energy level transitions. Sometimes atoms become ionized; they absorb so much energy that the electrons are stripped away, leaving positively charged nuclei. Lines resulting from ionization need to be accounted for. Molecules, composed of combinations of atoms, introduce *molecular bands* into the spectrum rather than discrete lines. Within molecules, atoms can vibrate in various ways, creating a "smeared" molecular band composed of numerous indistinguishable spectral lines. Particular molecules have characteristic bands.

How Are Atoms Excited?

It may appear from this discussion that electrons are raised to higher energy levels only by the absorption of electromagnetic energy. Actually, this is only one of three ways in which atoms become excited. A second situation involves heating. Obviously, raising the temperature of substances can cause them to radiate; the filament of a light bulb emits light when it is heated. Atoms and molecules in a substance are in rapid motion, and innumerable collisions are occurring among them. Temperature is a measure of the speed of this motion; when temperature is raised, the motion is increased and collisions become more frequent and energetic. Collisions due to this heat energy can raise electrons to higher energy levels. A third mechanism involves bombardment of atoms with electrons, as in the television picture tube. Electrons emitted in a stream at the rear of the tube are accelerated so that they strike the front of the tube. Electrons in atoms coating the face of the tube are raised to higher energy levels; they then emit light when they return to lower levels.

The Doppler Effect

The Doppler effect was alluded to in Chapter 3 in the discussions of radial velocity, spectroscopic binaries, and receding galaxies. Most of us have observed this phenomenon in connnection with sound waves, but it is equally applicable to light waves. The pitch of the sound of an automobile horn rises noticeably as the horn approaches, and falls distinctly after it has passed. As the horn approaches, more waves per second reach the ear. This higher frequency (in effect, shorter wavelength) results in higher-pitched sound. Conversely, the frequency of waves reaching the ear from the rapidly receding horn is lower, and this results in sounds of lower pitch. In the case of light waves, a change in frequency (wavelength) results in a change of color. If a source of light is rapidly approaching the spectroscope, the wavelengths of the light will be shortened and all of the spectral lines will be shifted toward the violet end of the spectrum. If the source of light is receding, the wavelengths will be lengthened and all of the spectral lines will be shifted toward the red end of the spectrum. The amount of shifting in spectral lines indicates the velocity of approach or recession. The Doppler effect is the means by which many celestial motions are detected and measured. (See Figures 5.13 and 5.14.)

Stellar Temperatures

The colors of stars are clues to their temperatures. When a piece of iron is heated to incandescence, it at first appears "red hot." As its temperature rises its appearance changes, until it is finally "white hot." Both

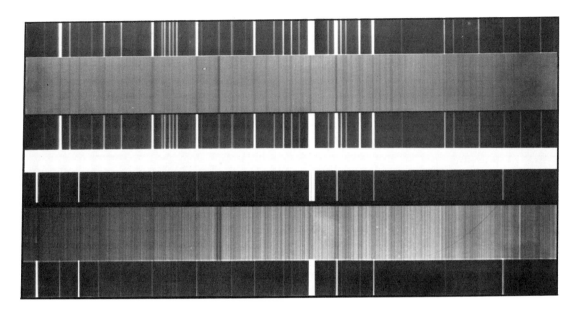

FIGURE 5.13 Two spectrograms of Zeta Ursae Majoris (Mizar), a spectrographic binary. Lines are double in one plate and single in the other. The double lines are produced when the motion of one star of the pair is toward the earth and that of the other is away from the earth. The single lines are produced when the motion of neither star is toward or away from the earth. The bright lines were not produced by light from Mizar; they are comparison lines added by the photographer to assist in measuring the displacement of Mizar's lines. (Yerkes Observatory photograph.)

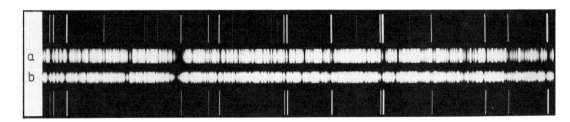

FIGURE 5.14 Two spectrograms of the star Arcturus taken about six months apart. The bright comparison lines reveal that the dark lines in the spectra of Arcturus are shifted toward the red in one photograph and toward the violet in the other. The difference in the velocities of approach and recession (about 50 km/sec) is due to the orbital velocity of the earth. (Photographs courtesy of the Hale Observatories.)

red-hot and white-hot iron produce continuous spectra, but there are still differences in the spectra. The intensities of particular regions of the spectra differ; the wavelengths toward the red end are relatively more intense in the spectrum of red-hot iron, whereas the wavelengths toward the violet end are more intense in the spectrum of white-hot iron. There are differences of this sort among stellar spectra. Plotting the relative intensity against the wavelength of radiation emitted by a body at a particular temperature can be very revealing. Some such plots, called *Planck curves,* are illustrated in Figure 5.15. The temperature indicated (T) is on the Kelvin or absolute scale. Zero on this scale

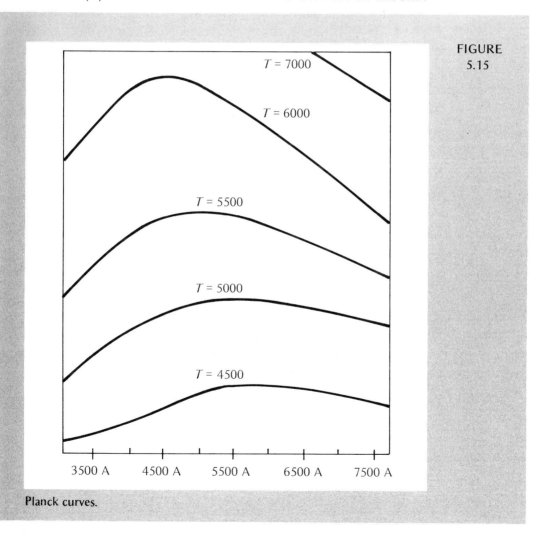

FIGURE
5.15

Planck curves.

is the temperature at which all molecular motion stops; it is the equivalent of −273 degrees on the centigrade or Celsius scale. A degree of temperature change on the Kelvin scale is equal to a degree of change on the Celsius scale. Planck curves reveal two important patterns: (1) every point on a Planck curve for a particular temperature is higher than every point on a curve for a lower temperature, and (2) the curves for progressively higher temperatures peak at increasingly shorter wavelengths.

Late in the last century, Wilhelm Wien derived a relationship between the wavelength of maximum intensity and the absolute temperature of a radiating body. This relationship is known as Wien's law:

$$T = \frac{3 \times 10^7}{\lambda_{max}}$$

The wavelength of maximum intensity in sunlight is about 5,000 A. Applying Wien's law, the surface temperature of the sun is

$$T = \frac{3 \times 10^7}{5,000}$$

$$T = 6,000 \text{ degrees}$$

Another relationship between temperature and radiation is the Stefan-Boltzmann law:

$$Q = AT^4$$

Q is equal to the radiation in ergs/sec from 1 cm² of surface. (An erg is the work done by a force of 10^{-5} N acting through a distance of 1 cm.) A is a constant equal to approximately 6×10^{-5}. We can compare the surface temperature of the sun calculated by this relationship to that calculated by Wien's law. The total surface area of the sun is about 6×10^{22} cm², and the total energy emitted by the sun is about 4×10^{33} ergs/sec. The energy emitted from one square centimeter of the sun's surface is then about 7×10^{10} ergs/sec. Thus,

$$Q = AT^4$$

$$T^4 = \frac{Q}{A}$$

$$T^4 = \frac{7 \times 10^{10}}{6 \times 10^{-5}}$$

$$T = 6,000 \text{ degrees}$$

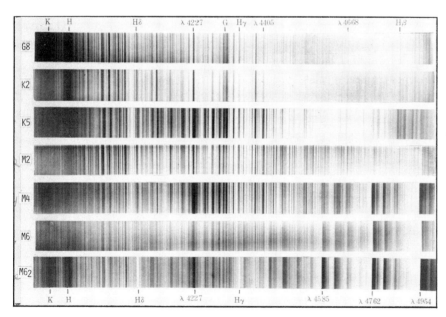

FIGURE
5.16

Representative stellar spectra in the violet. (Photograph courtesy of the Observatory of The University of Michigan.)

Spectral Classes

Spectrograms of a vast number of stars have been available for a long time. Part of the scientific process is the examination of data in search of revealing patterns. Astronomers have observed patterns among stellar spectra and have developed a classification scheme based upon them. An important contributor to this area of astronomy was Annie J. Cannon of Harvard University, who classified the spectra of nearly 400,000 stars. The scheme that she helped to develop recognizes seven principal spectral classes, which are designated O, B, A, F, G, K, M. The patterns of the absorption lines increase in complexity from class O to class K. Finer distinctions among spectra are indicated by breaking the principal classes into ten subclasses designated with numbers from 0 to 9. Thus, the spectral class of a star is generally identified by a letter and a number; the sun's is G2. Megh Nad Saha, an Indian astrophysicist, did important theoretical work in 1920 and 1921 in which he related the Bohr atom to spectral types. He theorized that a given element should experience particular amounts of ionization at particular temperatures. It follows from this that absorption lines of particular elements should be most prominent at particular temperatures. Thus the spectral class of a star reveals information about its temperature. Since spectral classes are related to temperature, they are also related to color. Blue, blue-white, and white stars have spectra of classes O, B, and K; yellow to orange stars are in classes F, G, and K; red stars have class M spectra. Figure 5.16 illustrates some differences among spectral classes.

MEASURING THE STARS

Stellar Diameters

How can the diameter of a star be determined when it appears as only a point of light in even the largest telescopes? Generally, the size of a star is inferred from its luminosity and temperature. But the angular diameters of a very few of the nearest, brightest, and largest stars have been determined using the technique of *interferometry.*

Interference of light waves was mentioned in connection with diffraction. Interferometry requires an arrangement of mirrors that directs two parallel beams of light into the telescope. Waves in the beams of light coming from two different regions of the star should theoretically interfere constructively and destructively to form a pattern of light and dark lines in the telescope eyepiece. By varying the distance between the mirrors to cause or to eliminate the interference pattern, the angular diameter of the star can be calculated from the geometry of the situation. If the star's distance is known, its diameter can then be calculated.

A star's size can be inferred from its temperature and brightness. Spectral class indicates temperature. The Stefan-Boltzmann law relates the amount of energy radiated per unit of surface area to temperature; that is, two stars of equal absolute temperature radiate equal quantities of energy from equal surface areas. Therefore, if two stars of equal temperature have unequal absolute magnitudes, the brighter star is larger. A magnitude difference of 1 represents a brightness difference of 2.512 times; a magnitude difference of 5 represents a brightness difference of 100 times. A star whose absolute magnitude is 5 more than that of another star of equal temperature has 100 times as much surface area as the other star. Formulas have been derived that relate absolute magnitude, temperature, and size. Thus, size can be calculated when temperature and magnitude are known.

The sizes of stars vary enormously. Some *red giants*, which are cool, red, and yet bright stars (absolute magnitudes of about −2), are so large that the orbit of Mars could be contained within them. Antares in Scorpius and Betelgeuse in Orion are examples of such stars. On the other hand, *white dwarfs*, which are white-hot faint stars (absolute magnitudes of about +12), are smaller than the earth. Pulsars may actually be less than 150 kilometers in diameter.

Stellar Masses

The mass of the earth can be calculated from its gravitational effect on falling bodies; the mass of the sun can then be calculated from its gravitational effect on the earth. The sun's mass is about 2×10^{33} grams. The masses of stars in binary pairs can sometimes be determined from relationships derived by Kepler and Newton. Using Kepler's third law and his own law of universal gravitation, Newton developed the relationship

$$P^2 (m_1 + m_2) = d^3$$

P represents the period in years, $(m_1 + m_2)$ represents the sum of the masses of an orbiting pair (such as binary stars) in units of the sun's mass, and d represents the distance between the objects in astronomical units. When astronomers know the distance between binary stars and their periods of revolution, they can compute the sum of their masses in units of solar mass. When they know the location of the center of mass of the pair, they can determine the masses of the individual stars, because the mass of one star times its distance from the center of mass equals the mass of the other star times its distance from the center of mass.

Plotting the masses against the absolute magnitudes of the stars for which both of these values are known (about 100) reveals a rather consistent relationship; the more massive stars are the more luminous. (See Figure 5.17.) This curve is used to estimate the masses of stars for which only the absolute magnitude is known.

FIGURE 5.17

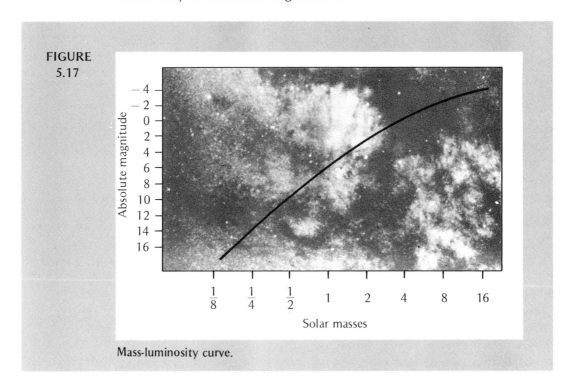

Mass-luminosity curve.

Masses of stars vary widely, but relatively speaking the range of masses is less than the range of sizes. The lightest stars have masses somewhat less than 10% of the sun's mass, and the masses of the heaviest stars approach 100 times the solar mass.

When size and mass are known, stellar density can be calculated. The extremes of stellar densities are particularly striking. Some of the red supergiants are especially tenuous, with densities less than 1/1000 that of air. In contrast, a spoonful of material from some of the white dwarf stars would weigh several tons, and a spoonful of material from neutron stars would weigh thousands of tons. The sun's average density is a little less than one and one-half times that of water.

Stellar Distances

Relatively few stars are close enough to allow us to measure their distances by parallactic techniques. The difficulties involved in detecting parallactic displacement of even the closest stars have been discussed. Fortunately, the distances to many stars can be found from a relationship of absolute magnitude, apparent magnitude, and distance.

The brightness of a light appears to diminish as it is moved away. The intensity of illumination reaching an observer varies inversely with the square of the distance to the source of light. For instance, the intensity of light from a 100-watt bulb 10 meters from an observer will diminish to one-ninth that intensity if the bulb is moved to a distance of 30 meters. It follows that you could determine your distance from a 100-watt bulb by using a light meter to measure the intensity of illumination reaching you and comparing that to the intensity of illumination from a 100-watt bulb at a known distance. Similarly, astronomers can determine the distance to a star by comparing its absolute magnitude (its brightness at a distance of 10 parsecs) to its apparent magnitude (its brightness as viewed from the unknown distance). Based on the inverse square relationship between distance and intensity of illumination, astronomers have derived an equation that relates absolute magnitude, apparent magnitude, and distance. The equation is

$$M = m + 5 - 5 \log r$$

M is absolute magnitude, m is apparent magnitude, and $\log r$ is the logarithm of the distance in parsecs. But is it conceivable that absolute magnitude could be known before distance? It is in the case of some kinds of variable stars, including Cepheid variables. Cepheids are recognized by the shape of their light curves—the curve formed by plotting magnitude against time (Figure 5.18). There are several hundred known Cepheids in our Milky Way Galaxy. The North Star, Polaris, is a Cepheid. Since Cepheids are very luminous, some can be seen in other galaxies.

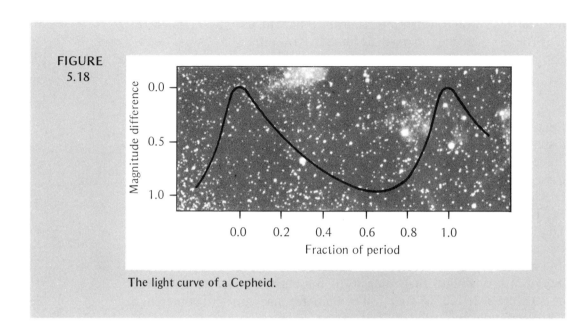

FIGURE
5.18

The light curve of a Cepheid.

In 1912, Henrietta Leavitt of Harvard University was studying stars of the Magellanic Clouds. The Magellanic Clouds are two small galaxies that can be seen about 20 degrees from the south celestial pole. They are the closest galaxies to our Milky Way Galaxy; they are at a distance of about 170,000 light years. Their name comes from the fact that they were described by sailors who had accompanied Magellan on his ill-fated voyage to the Philippine Islands. Leavitt noted several variable stars (later recognized as Cepheids) in the Magellanic Clouds and observed that the brightest stars had the longest periods. When she plotted average apparent magnitude against period, she found a very regular relationship; the plotted points fell close to the curve indicated in Figure 5.19. She realized that all of these Cepheids were at essentially the same distance from the earth, for distances between stars are minor compared to the distance from the earth to the Magellanic Clouds.

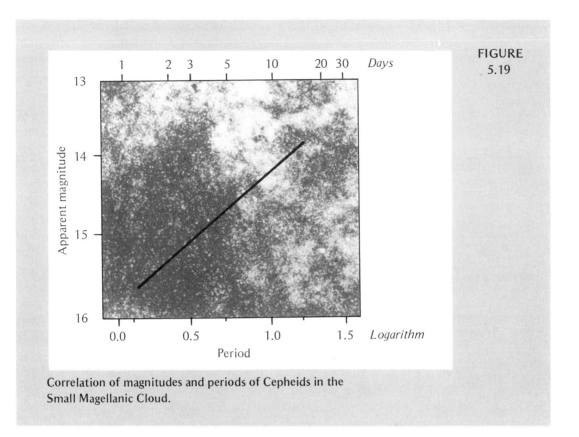

FIGURE
5.19

Correlation of magnitudes and periods of Cepheids in the
Small Magellanic Cloud.

Hence, their absolute magnitudes, as well as their apparent magnitudes, varied with period. It was clear that if the distance to a single Cepheid could be determined, its absolute magnitude could be calculated, and the period-luminosity curve could be calibrated. Once the curve was calibrated, the absolute magnitude of any Cepheid would be evident from its period. Then, knowing both absolute and apparent magnitudes, distance to the Cepheid could be determined.

It was difficult to calibrate the curve; no Cepheid was close enough to the sun for a direct measurement of its parallax. By 1918, Ejnar Hertzsprung in Denmark and Harlow Shapley at Mt. Wilson Observatory had calibrated the period-luminosity relation by means of a statistical analysis of the motions of a dozen Milky Way Cepheids. They reasoned that, on the average, the smaller the mean distance of a star in this stellar sample, the greater its proper motion would be. If this were the case, the tangential velocity corresponding to this proper motion could be estimated with the aid of the measured radial velocities of the stars, assuming an average relationship between radial and tangential velocities. We know now that, apart from the small size of the sample and the inaccuracies in their measured motions, there were two major sources of error in the calibration. Galactic rotation and the dimming influence of interstellar dust in and near the plane of the Milky Way were not understood at this time, and hence appropriate corrections were not made. As a result, the estimates by Hertzsprung and Shapley of the absolute magnitudes of the Milky Way Cepheids were 1.5 magnitudes too faint.

Soon after his work on the calibration of the Cepheids, Shapley also determined the absolute magnitudes of another useful class of variables now called RR Lyrae stars. These vary in regular periods of less than one day, and all have the same absolute magnitude regardless of their period. They have been found in large numbers in some of the globular clusters that surround the disk of the Galaxy like a halo. In these clusters there are also a few variables with light curves much like those of the Milky Way Cepheids (now called Population I Cepheids) that Shapley had been studying, but (as was later discovered) with absolute magnitudes 1.5 magnitudes fainter. Shapley found that the RR Lyrae variables in a globular cluster always fell on a smooth extension of the period-luminosity relation for the cluster Cepheids. Assuming for the latter the erroneous calibration for Population I Cepheids, he arrived, by a happy cancelling of two errors, at a fairly accurate absolute magnitude for the RR Lyrae variables. Everything seemed to be in order. In 1925, Edwin Hubble, using the 100-inch telescope at Mt. Wilson and Shapley's period-luminosity calibration for Cepheid variables, determined the distances to two nearby spiral galaxies (one was the spiral M31 in Andromeda).

It was not until 1952 that Walter F. Baade, working with the 200-inch reflector at Mt. Palomar, showed that Shapley's assumption that RR Lyrae variables and Milky Way (Population I) Cepheids lay on the same period-luminosity curve was inconsistent with Baade's observations of M31. Baade concluded that the Milky Way Cepheids and those in M31 were 1.5 magnitudes brighter than hitherto supposed and that the Cepheids in globular clusters were a different type of star (now called Population II Cepheids). Almost overnight, astronomers realized that the distances to M31 and other nearby galaxies were double those previously adopted, and the whole scale of the universe doubled in size!

Despite problems getting the scale right, the period-luminosity curve has been immensely useful in determining stellar distances. In addition to techniques involving variable stars, there are other methods for determining distance that also involve comparisons of apparent and assumed intrinsic luminosities of various kinds of objects.

It is possible to beam radio and light waves to the nearest solar system objects and to detect some of this radiation reflected to the earth. With radar, the radiation is radio waves; with lasers, the radiation

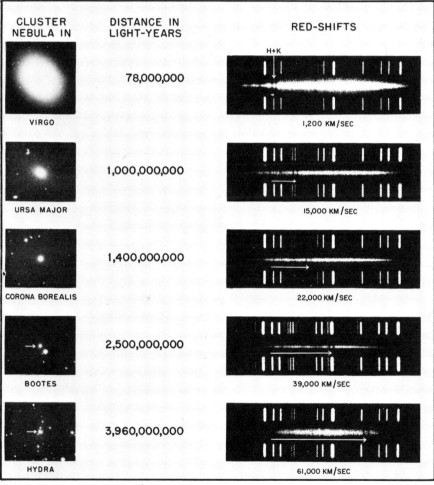

FIGURE
5.20

CLUSTER NEBULA IN	DISTANCE IN LIGHT-YEARS	RED-SHIFTS
VIRGO	78,000,000	H+K → 1,200 KM/SEC
URSA MAJOR	1,000,000,000	15,000 KM/SEC
CORONA BOREALIS	1,400,000,000	22,000 KM/SEC
BOOTES	2,500,000,000	39,000 KM/SEC
HYDRA	3,960,000,000	61,000 KM/SEC

Relation between red shift (velocity) and distance for distant galaxies. Arrows indicate the shift of two calcium lines.
(Photograph courtesy of the Hale Observatories.)

is light. These techniques are particularly useful in the determination of distances, which can be calculated from the time required for the round trip and the velocity of the radiation. It is also possible to make some conjectures about the properties of the surfaces that reflected the radiation. The availability of these new techniques creates some important research possibilities.

The technique for estimating distances to the most remote galaxies utilizes the red shift. Absorption lines in the spectra of all but a few of the closest galaxies are shifted toward the red end. The amount of red shift increases with diminishing brightness. It appears that a relationship exists between the speed of recession and distance; the farther away a galaxy is, the greater its speed of recession and the greater the red shift. The amount of red shift therefore becomes a means of estimating distance, although astronomers do not have great confidence in the estimated distances to the farther galaxies. (See Figure 5.20.)

OPTICAL TELESCOPES

Lenses

If proper consideration is given to its shape and index of refraction, a lens can be designed to redirect the path of light in any way desired. A convex lens can be shaped so that all of the light rays passing through it will converge and pass through a single point called the *focal point,* as in Figure 5.21. (The use of a lens as a burning glass illustrates this phenomenon.) The distance from the center of the lens to the focal point is the *focal length*. Actually, a lens made of a single piece of glass will only approximately refract all the light through a focal point. Since the glass has different indices of refraction for light of different wavelengths, light of each color will have its own focal point. If the lens is to be used in an optical instrument, these differing focal points cause an undesirable effect known as *chromatic aberration.* Chromatic aberration can be minimized by using two pieces of glass of differing indices of refraction shaped to fit together and act as a single *achromatic lens.* (See Figure 5.22.)

The light rays that have passed through a lens can form an optical image of the object from which they have come. This is easily demon-

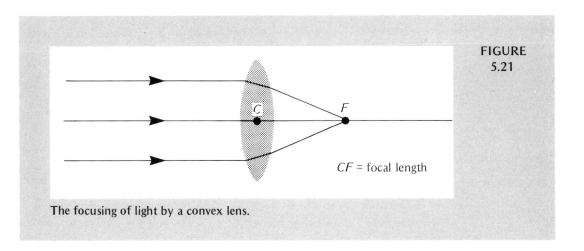

FIGURE
5.21

CF = focal length

The focusing of light by a convex lens.

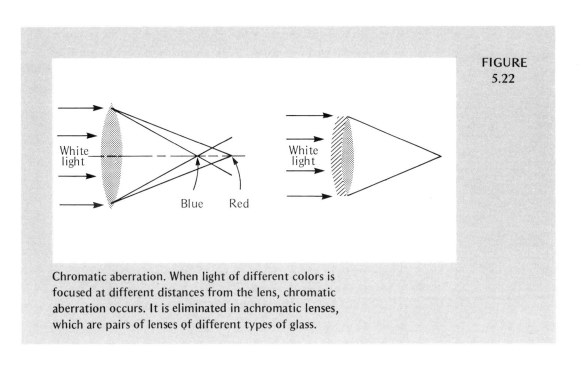

FIGURE
5.22

Chromatic aberration. When light of different colors is
focused at different distances from the lens, chromatic
aberration occurs. It is eliminated in achromatic lenses,
which are pairs of lenses of different types of glass.

strated by using a sheet of paper as a projection screen; a convex lens
held between an object and the paper will cast an image of the object
onto the paper. Figure 5.23 diagrams such image formation; notice that
the image is inverted.

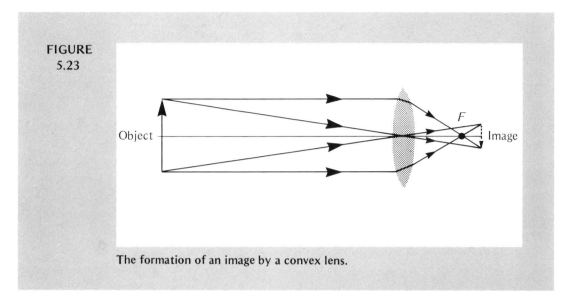

The formation of an image by a convex lens.

A lens can serve as a magnifying glass if one looks through it in such a way that the object viewed and its image are on the same side of the lens. This effect is diagrammed in Figure 5.24. Note that the image is erect, not inverted.

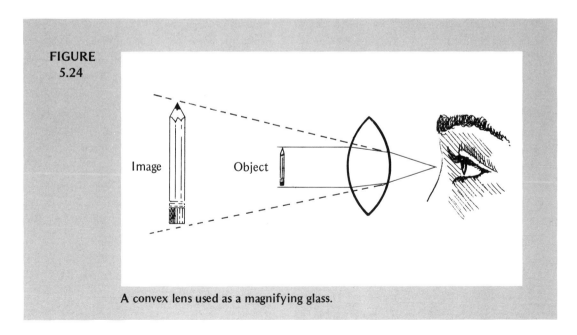

A convex lens used as a magnifying glass.

Telescopes

Both ways of forming images are used in a telescope. A lens is used to produce an inverted image, and then a second lens is used as a magnifying glass through which we view the image formed by the first lens. This arrangement is diagrammed in Figure 5.25. Note that the image seen is inverted. This is not inconvenient in an astronomical telescope. In a telescope used for terrestrial viewing, a third lens is placed between the other two. The purpose of this lens is to re-invert the image formed by the first lens, so that the image finally viewed is right-side up.

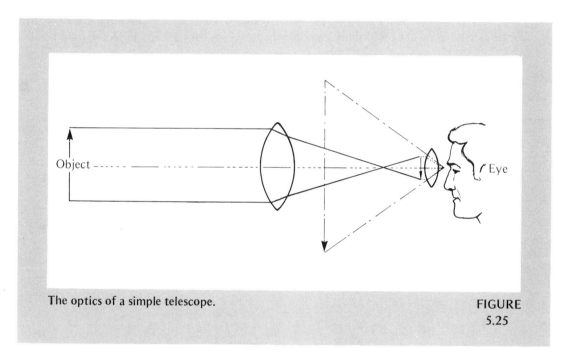

The optics of a simple telescope.

FIGURE
5.25

Telescopes utilizing combinations of lenses are called *refractors.* The larger lens, which gathers light to form the initial image, is the *objective* lens. The lens next to the eye, which is used as a magnifying glass, is the *eyepiece.* The *magnifying power* or *magnification* of a telescope is equal to the focal length of the objective divided by the focal length of the eyepiece. A telescope that has an objective with a focal length of 800 millimeters and an eyepiece with a focal length of 4 millimeters magnifies 200 times. Unless a bright image of high quality has been formed by the objective, it is futile to magnify it with the eyepiece. It is seldom useful to use an eyepiece that results in magnification of more than 40 times the diameter of the objective measured in inches. A particularly critical quality of a telescope is its *resolving power.* Resolving power refers to the minimum angular diameter an object must have in order to be seen with the telescope; it indicates the minimum angular separation required between two stars before they will appear as two points of light in the telescope rather than merging into a single, fuzzy point of light. The resolving power of the unaided eye is about 6 minutes of arc. The resolving power of a telescope is limited by the wavelengths of the radiation and the diameter of the objective. Resolving power in seconds of arc is roughly equal to 5 divided by the diameter of the objective in inches. A telescope with a 10-inch objective has a theoretical resolving power of about 0.5 second. Actually, the resolving power of telescopes is limited by the *seeing* to about two seconds of arc. Seeing refers to the quality of the viewing as influenced by light refraction from the turbulence and varying density of the atmosphere. Seeing causes such effects as the twinkling of stars. Bad seeing at the earth's surface makes astronomers eager for space-age opportunities to lift telescopes above the atmosphere.

Large refracting telescopes are particularly difficult to make for several reasons. The glass must be of unusually high quality, since light must pass through the lenses. Construction of an achromatic objective requires the shaping of four surfaces. Large objectives are heavy and bend slightly due to their weight; this distorts the image. The largest refractor is at Yerkes Observatory in Wisconsin. Constructed in 1897, it has a 40-inch objective with a focal length of 63 feet.

In a *reflecting telescope,* a concave (usually parabolic) mirror gathers the light to form an image. The largest optical telescopes are reflectors. It is simpler to construct reflectors than refractors for several reasons. The weight of the mirror can be supported from behind. Only one surface needs to be shaped, since mirrors do not cause chromatic aberration. The material in the mirror need not be transparent; its purpose is to provide a surface that can be shaped and then coated with a thin, reflective layer of aluminum. The point of observation of a reflector is usually not at the focal point of the mirror. The light rays are generally directed to a more convenient location with a secondary mirror. Sometimes there is a hole through the center of the mirror so

FIGURE
5.26

The 36-inch refracting telescope of Lick Observatory. (Lick
Observatory photograph.)

that the eyepiece can be used behind the mirror. Two common arrange-
ments are illustrated in Figure 5.27. Since 1948, the largest American
reflector has been the 200-inch Hale telescope on Mt. Palomar in
California. There is a much newer 236-inch reflector in Russia.

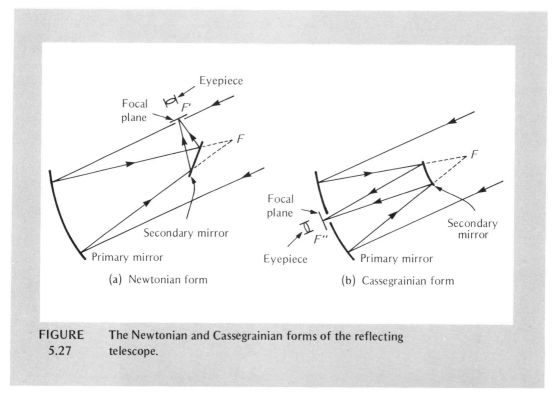

FIGURE
5.27
The Newtonian and Cassegrainian forms of the reflecting telescope.

FIGURE
5.28

The 200-inch mirror of the Hale telescope. There is an observer in a cage at the focal point. (Photograph courtesy of the Hale Observatories.)

FIGURE
5.29

The 120-inch reflecting telescope of Lick Observatory.
(Lick Observatory photograph.)

Newer in design than refractors and reflectors are some telescopes that combine features of each. These are a form of reflecting telescope in which a lens is located in the position occupied by the objective in a refractor. This lens does not form an image. Rather, it works in combination with the mirror by slightly altering the paths of the rays before they strike the mirror. This combination of *correcting plate* and mirror (usually spherical) can provide some results not obtainable from mirrors alone. One such lens-mirror system is the *Schmidt telescope,* which is particularly useful for wide-angle astronomical photography. (See Figures 5.30 and 5.31.)

FIGURE 5.30

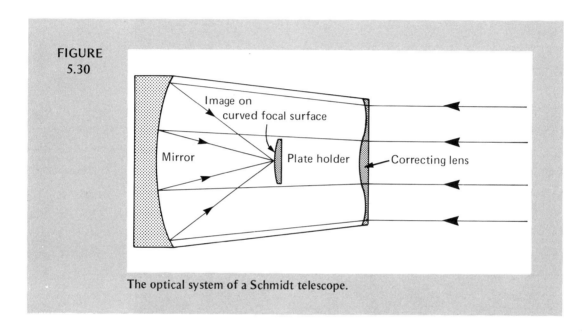

The optical system of a Schmidt telescope.

Astrophotography

Astronomers seldom look through major telescopes. In general, telescopes are used to photograph the sky; the eyepiece is replaced by a photographic plate. This has two obvious advantages: (1) photographs are permanent records that can be studied anytime, and (2) time-exposure photographs can reveal objects and details too faint for the eye to see. In addition, photographic emulsions can be prepared to be especially sensitive to particular wavelengths. Careful selection of emulsions and filters results in photographs that reveal useful information about star colors.

FIGURE
5.31

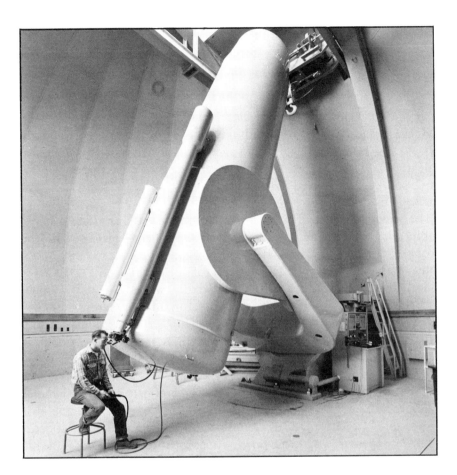

The Schmidt telescope on Mt. Palomar. The correcting lens
has a diameter of 48 inches; the diameter of the mirror is
72 inches. (Photograph courtesy of the Hale Observatories.)

Image Tubes

The image tube is an electronic device for intensifying the image
produced by a telescope objective or mirror. The image is focused on a
light-sensitive surface from which it dislodges electric charges—the
photoelectric effect. These charges are given an acceleration and allowed to strike a fluorescent screen. The image formed on the screen
replicates with increased intensity the image formed by the mirror or
objective. The screen is then photographed.

197

OBSERVING WITH INVISIBLE RADIATION

Radio Telescopes

The first detection of radio waves from space was a surprise. In the early 1930s, Karl Jansky, a radio engineer for the Bell Telephone Laboratories, was studying atmospheric disturbances in radio communications. He found that his radio receiver was detecting emissions from a source that moved as the sun crossed the sky. This source was the sun.

The electromagnetic radio waves from celestial sources can be focused with concave reflectors just as visible light is focused by the mirrors of optical telescopes. In radio telescopes, eyepieces and photographic plates are replaced by devices for measuring the wavelengths and intensities of radio waves. Because radio waves are vastly longer than light waves, the surfaces of the reflectors in radio telescopes need not be smoothed as precisely as those in optical telescopes. The reflectors are sometimes made of wire mesh. It is difficult to achieve adequate resolving power in radio telescopes, because the long wavelengths require reflectors of great diameter. The Jodrell Bank radio telescope in England has a steerable, parabolic reflector with a diameter of 250 feet. Some telescopes, which are either partially steerable or non-movable, are larger. In Arecibo, Puerto Rico, a 1,000-foot reflector has been shaped in a natural bowl. As substitutes for huge, movable reflectors, very large arrays of manageably small reflectors can work in combination.

21-cm Radio Waves

The possibility of using radio astronomy as a means of studying tenuous interstellar material was mentioned in Chapter 3. Hydrogen atoms are more abundant than other kinds of atoms in stars, and it was predicted early in this century that the space between stars would be found to be pervaded by very tenuous gaseous hydrogen. There was no way to detect it, because Bohr model considerations indicated that light absorbed by the hydrogen would be of wavelengths too short to cause absorption lines in the visible spectrum. It was predicted in 1945 that interstellar hydrogen emitted 21-cm radio waves; such radiation was

The 140-foot telescope of the National Radio Astronomy Observatory, Green Bank, West Virginia. (Photograph courtesy of the National Radio Astronomy Observatory.)

FIGURE
5.32

detected in 1951. Study of the distribution of hydrogen, using its 21-cm radiation, has been immensely useful in mapping the spiral arms in our Milky Way galaxy.

The emission of a 21-cm radio wave does not result from an electron dropping into an orbit closer to the nucleus. In the typical interstellar hydrogen atom, the electron is already in the orbit closest to the nucleus; the atom is said to be in the ground state. The proton at the center of the atom is spinning, and the electron is also spinning as it orbits the proton. Since they are moving electric charges, they generate magnetic fields. The energy of these magnetic fields contributes to the total energy of the atom. If the electron rotates with its magnetic axis

FIGURE
5.33

The 1,000-foot radio telescope in Arecibo, Puerto Rico.
(Cornell University photograph, operator of the Arecibo
Observatory under contract with the National Science
Foundation.)

parallel to the proton, it is in a higher energy state than if it flips
end-over-end. When the axis of an electron in the higher energy state
flips, a photon of 21-cm wavelength is emitted.

The Newest View of the Sky

Astronomers now have instruments that permit them to use invisible
gamma rays, x-rays, ultraviolet waves, and infrared waves to study the

sky. The first observations using invisible radiation were in radio astronomy. Unanswered questions about pulsars and quasars and the recent detection of radio waves from additional kinds of interstellar molecules demonstrate that the work of the radio astronomer is far from done. But newer technology which makes it possible to work with shorter-wavelength invisible radiation has opened additional research frontiers.

Not only are the sky's ultraviolet waves, x-rays, and gamma rays invisible, but they are not available for study at the earth's surface. The atmosphere absorbs these wavelengths and shields the surface from them. Balloons, rockets, and orbiting artificial satellites can now raise instruments above the shielding atmosphere. The technological problems involved in recording the invisible radiation and transmitting it back to the earth's surface are complex. Although photographic film can be made sensitive to wavelengths invisible to the eye, problems involved in recovering the film and other considerations usually require a different approach. A variety of instruments have been developed for detecting radiation of particular wavelengths and transmitting information about it by radio waves to the earth's surface, where the characteristics of the radiation can be reconstructed from the varying intensity of the radio signal. Observations in ultraviolet, x-ray, and gamma-ray astronomy are being used in mapping the sky at these wavelengths. When completed, these observations will indicate the distribution of the sources of such radiation. Identification of specific sources and analyses of the radiation from them will follow; indeed, this work has already begun. The ultraviolet spectrum of the sun has been studied in considerable detail. More than 100 x-ray sources have been identified, as well as a general background of x-rays that may be coming from beyond our Galaxy. Several specific sources of gamma rays have been detected, along with diffuse radiation coming from the direction of our Galaxy's center.

Considerable infrared radiation reaches the earth's surface, and instruments do not need to be above the atmosphere to detect it. In recent years a few large telescopes have been built especially for infrared observation. The technology is complex. Almost everything emits infrared wavelengths, because everything contains heat. The astronomer must identify the infrared radiation of a star in the midst of all this other radiation. Several thousand infrared sources have been identified. Stars brightest in the infrared wavelengths are not always brightest in visible wavelengths. If our eyes were sensitive to another band of wavelengths, the sky would appear quite different.

POLLUTED SKIES

Astronomers share with biologists, ecologists, and others a professional concern for the environment. Astronomers are alarmed about the deteriorating physical environment. Major observatories were built on mountains or other elevations and away from cities, but urban sprawl has brought a befouled atmosphere and lighted skies close to observatories that seemed safely removed when constructed. Particularly distressing to astronomers are city lights, which can fog their photographic plates before they can record faint objects.

CONCLUSION

Gravitation has confined us to the earth until very recently; we are still able to venture only a relatively short distance into space. Accordingly, we have had to learn about the universe by studying the stream of radiant energy in which our planet is bathed. It was difficult for us to come to understand the nature and generation of electromagnetic energy, even though it is this energy from the sun that sustains life. Now we have an array of instruments to receive and facilitate our interpretation of this energy: cameras, telescopes, spectroscopes, radio receivers, etc. The radiant energy from distant stars is so feeble when it reaches the earth that we cannot detect it without sensitive instruments, but this energy tells us much about the distances, sizes, masses, composition, temperatures, and motions of the stars.

FOR REVIEW, DISCUSSION, OR FURTHER STUDY

1. Describe one effect of light that suggests that light is a stream of particles and another effect that suggests that it is a wave phenomenon.

2. If a boulder is dropped into a pond, waves moving outward can rock a rowboat several feet away. The water uses waves to transmit the energy from the falling stone to the boat. If light energy is transmitted through the vacuum of space, what is moving? If there is nothing to move, how can there be waves?

3. What was the nineteenth-century ether hypothesis?

4. What was the conflict between Newtonian mechanics and electromagnetic theory that the Bohr atom seemed to resolve?

5. Do bright nebulae produce bright-line or dark-line spectra?

6. Betelgeuse in Orion is at a distance of 520 light years, has an apparent or visual magnitude of about 0.4, and is a relatively cool star of spectral type M2. Mizar, a star in the handle of the Big Dipper, is at a distance of 88 light years, has an apparent magnitude of 2.25, and is a relatively hot star of spectral type A2. If Betelgeuse is cooler and much more distant than Mizar, why is it brighter?

7. Why are the largest optical telescopes reflectors rather than refractors?

8. Who is Grote Reber?

9. How many times greater is the light-gathering power of a 200-inch telescope than that of a 2-inch telescope?

10. Describe and explain the phenomenon astronomers refer to as "red shift." How is this phenomenon utilized?

SUGGESTED ACTIVITIES

A Phonograph Record as a Diffraction Grating

You can demonstrate the ability of a grating to produce a spectrum by holding a long-playing phonograph record in sunlight. Hold the record at arm's length so that it reflects sunlight into your eyes. Tilt the record so that the light is reflected at various angles. In some positions you will see a spectrum against the surface of the record; the record is functioning as a reflection grating.

Creating a Telescope

Obtain two convex lenses, such as a reading glass and a pocket magnifying glass. They should be of different focal lengths. You can determine whether or not they are of different focal lengths by seeing whether they focus images at different distances. Hold a lens in one hand and a sheet of paper in the other in such a way that light from a distant object (such as a tree several feet away) passes through the lens onto the paper. Move the lens toward or away from the paper until a clear image of the object is formed on the paper. Note the distance of the lens from the paper, and note that the image is inverted and minified. Hold the lens of longer focal length with one hand and that of shorter focal length close to the eye with the other hand. Look through the two lenses, positioning them so that the image formed by the lens of longer focal length (objective lens) is magnified by the lens held close to the eye (the eyepiece).

SUGGESTED READINGS

Charles, Philip A., and J. Leonard Culhane, "X Rays from Supernova Remnants," *Scientific American*, Vol. 233, No. 6 (December 1975), 38.

Fichtel, Carl, Kenneth Greisen, and Donald Kniffen, "High-Energy Gamma-Ray Astronomy," *Physics Today*, Vol. 28, No. 9 (September 1975), 42.

Friedman, Herbert, "Cosmic X-ray Sources: A Progress Report," *Science*, Vol. 181 (3 August 1973), 395.

Goldberg, Leo, "Ultraviolet Astronomy," *Scientific American*, Vol. 220, No. 6 (June 1969), 92.

Goldhaber, Alfred Scharff, and Michael Martin Nieto, "The Mass of the Photons," *Scientific American*, Vol. 234, No. 5 (May 1976), 86.

Gursky, Herbert, and Edward P. J. van den Heunel, "X-Ray–Emitting Double Stars," *Scientific American*, Vol. 232, No. 3 (March 1975), 24.

Hammond, Allen L., "Laser Ranging: Measuring the Moon's Distance," *Science*, Vol. 170 (18 December 1970), 1289.

Hammond, Allen L., "Ultraviolet Astronomy: Progress with the OAO," *Science*, Vol. 170 (27 November 1970), 960.

Hjellming, R. M., "An Astronomical Puzzle Called Cygnus X-3," *Science*, Vol. 182 (14 December 1973), 1089.

Israel, Martin H., P. Buford Price, and C. Jake Waddington, "Ultraviolet Cosmic Rays," *Physics Today*, Vol. 28, No. 5 (May 1975), 23.

Kellerman, Kenneth I., "Extragalactic Radio Sources," *Physics Today*, Vol. 26, No. 10 (October 1973), 38.

LaLonde, L. M., "The Upgraded Arecibo Observatory," *Science*, Vol. 186 (18 October 1974), 213.

Metz, William D., "Astronomy: TV Cameras Are Replacing Photographic Plates," *Science*, Vol. 175 (31 March 1972), 1448.

Metz, William D., "Astronomy from an X-ray Satellite: Measuring the Mass of a Neutron Star," *Science*, Vol. 179 (2 March 1973), 884.

Metz, William D., "Astronomy from Space: New Class of X-ray Sources Found," *Science*, Vol. 189 (26 September 1975), 1073.

Metz, William D., "Gamma Rays: From Neutron Stars, Supernovas, or 'Beebees'?" *Science*, Vol. 182 (21 December 1973), 1234.

Metz, William D., "X-ray Astronomy (II): A New Breed of Pulsars," *Science*, Vol. 179 (9 March 1973), 986.

Metz, William D., "X-ray Astronomy (III): Searching for a Black Hole," *Science*, Vol. 179 (16 March 1973), 1113.

Neugebauer, G., and Eric E. Becklin, "The Brightest Infrared Sources," *Scientific American*, Vol. 228, No. 4 (April 1973), 28.

Neugebauer, G., and Robert L. Leighton, "The Infrared Sky," *Scientific American*, Vol. 219, No. 2 (August 1968), 51.

Riegel, Kurt W., "Light Pollution," *Science*, Vol. 179 (30 March 1973), 1285.

Ryle, M., "Radio Telescopes of Large Resolving Power," *Science*, Vol. 188 (13 June 1975), 1071.

A lecture delivered by Professor Ryle when he received the Nobel Prize in Physics, which he shared with Professor Hewish.

Sandage, Allen R., "The Red-Shift," *Scientific American*, Vol. 195, No. 3 (September 1956), 170.

Schnopper, Herbert W., and John P. Delvaille, "The X-ray Sky," *Scientific American*, Vol. 227, No. 1 (July 1972), 27.

Thomsen, Dietrick E., "Gravity-Wave Astronomy," *Science News*, Vol. 108 (23 and 30 August 1975), 136.

Thomsen, Dietrick E., "Stars of Sky and Screen," *Science News*, Vol. 108 (23 and 30 August 1975), 132.

This brief article includes a color photograph of Betelgeuse produced through a new speckle interferometry technique which revealed, for the first time, surface detail of a star other than the sun.

Thomsen, Dietrick E., "Toward the Diffraction Limit," *Science News*, Vol. 108 (23 and 30 August 1975), 138.

Webster, Adrian, "The Cosmic Background Radiation," *Scientific American*, Vol. 231, No. 2 (August 1974), 26.

FIGURE
6.1
Rosette Nebula in Monoceros: NGC 2237. The dark
globules may be areas in which protostars will form.
(Photograph courtesy of the Hale Observatories.)

STELLAR EVOLUTION:
Birth and Death
in the Cosmos

6

We have seen that astronomers can determine an impressive amount about a star that appears as a point of light in their telescopes. They can determine its distance, size, composition, mass, temperature, and motion. But is it conceivable that they might be able to learn something about the age of the point of light and where it came from? How did the stars come into existence? Is "creation" still occurring? Can a star shine eternally? How are galaxies formed? In order to consider such questions as these, we must discuss some aspects of the nuclear processes that occur within stars.

THE EQUIVALENCY OF MASS AND ENERGY

Progress toward understanding atomic structure and nuclear processes has been dramatic. One area of this progress has been in the development of theoretical explanations for the release of stellar energy. These explanations answered long-standing questions about the sun, our most familiar star. Early ideas that the sun was simply burning were abandoned about a century ago, when calculations of solar mass and energy output indicated that any known fuel would have been completely burned within the period of recorded history. A suggestion made by Herman von Helmholtz in 1853 was accepted widely for several years.

He believed that the sun's energy resulted from contraction due to the sun's own gravitation. He suggested that the outer portions of the sun were falling inward and that the energy of the matter moving inward was converted into heat and light through collisions among its particles. His theory was abandoned when geological fossil records suggested that there has been life on the earth for a longer time than Helmholtz's theory could accommodate.

$E = mc^2$ is perhaps the most famous of all equations. Albert Einstein postulated this relationship in 1905, the year in which he explained the photoelectric effect and published his theory of relativity. If E represents energy in units of ergs, m represents mass in grams and c represents the speed of light in centimeters per second (about 3×10^{10} cm/sec). The equation helps to explain how nuclear reactions can supply much of the raging outpouring of stellar energy. It expresses the equivalence between mass and energy; Einstein stated that the mass of a body is a measure of its energy content. Since the speed of light squared is an enormous figure, it is obvious that a small amount of mass is equivalent to a great deal of energy. The energy in one ounce of matter is sufficient to raise the temperature of several million tons of water from the freezing point to the boiling point. The reduction of mass in chemical reactions where energy is released (the burning of wood, for instance) is so minute that ordinarily it cannot be detected. In nuclear reactions there is a measurable reduction of mass.

HYDROGEN-HELIUM FUSION

The nuclei of stellar atoms react with one another in numerous ways to form more complex atoms from simple ones. The predominant and most fundamental kind of change is one in which atoms of hydrogen fuse to form atoms of helium, with the conversion of some mass into energy. Hydrogen atoms are by far the most abundant atoms throughout the universe. Two specific mechanisms have been formulated to describe and account for this hydrogen fusion, the *proton-proton cycle* and the *carbon-nitrogen cycle*. Although both processes occur in the sun, the proton-proton cycle is the dominant one. In some more massive stars, the carbon-nitrogen cycle dominates. Hans Bethe first explained the carbon-nitrogen cycle in 1939; later he and others de-

FIGURE
6.2

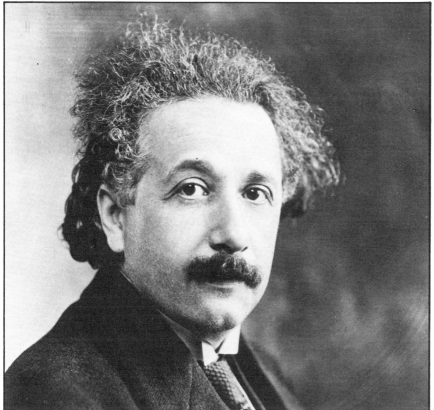

Albert Einstein. (Yerkes Observatory photograph.)

scribed the proton-proton cycle. Both of these cycles are often referred to as *hydrogen burning*, but the processes are far different from the chemical reactions of ordinary burning.

The essence of both cycles is that four hydrogen atoms yield one helium atom, gamma rays, and some neutrinos. Neutrinos are a kind of atomic particle without electric charge and very nearly without mass. The gamma rays cause the excitation of atoms and the torrent of electromagnetic energy described in the last chapter. The mass of the helium atom is about 1% less than the combined masses of the four hydrogen atoms. Both cycles proceed in a series of several reactions.

In order to describe the steps in these cycles, we will need to make use of some special notation. H and He are the symbols for hydrogen and helium, respectively. Subscripts indicate electric charges; superscripts indicate the mass number, which is the total number of protons and neutrons in the nucleus. Thus, the nucleus of an ordinary hydrogen atom (a proton) is $_1H^1$, and the nucleus of an ordinary helium atom is $_2He^4$. Helium's mass number of four results from two protons and 2 neutrons. The positron is an atomic particle with an electric charge of +1 and a mass equal to that of an electron; it is the counterpart of an electron.

The proton-proton cycle begins with a collision between the nuclei of two ionized hydrogen atoms (protons). Since they are positively charged, there is a natural force of repulsion between them and collisions ordinarily do not occur. For the cycle to proceed, two protons must collide so violently that they remain bound together. Such collisions do occur in the interior of the sun, where atoms have the violent motions associated with temperatures of 14,000,000 degrees Kelvin. Even there such collisions are rare; a given proton is likely to go for more than a billion years without experiencing such a collision. The proton-proton collision produces the nucleus of a heavy isotope of hydrogen (deuterium) plus a positron and a neutrino:

$$_1H^1 + {_1H^1} \rightarrow {_1H^2} + \text{positron} + \text{neutrino}$$

The neutrino escapes from the sun. The positron collides with an electron, and they annihilate one another to produce a gamma-ray photon:

$$\text{positron} + \text{electron} \rightarrow \text{gamma ray}$$

The deuterium nucleus (a deuteron) collides within seconds with another proton. This collision produces the nucleus of a light isotope of helium plus a gamma-ray photon:

$$_1H^2 + {_1H^1} \rightarrow {_2He^3} + \text{gamma ray}$$

Within a few hundred thousand years, on the average, the light helium nucleus collides with another light helium nucleus. This collision produces an ordinary helium nucleus plus two protons and another gamma-ray photon:

$$_2He^3 + {_2He^3} \rightarrow {_2He^4} + {_1H^1} + {_1H^1} + \text{gamma ray}$$

The cycle ends with this collision between two light helium nuclei. The interaction of six protons and two electrons has produced one helium nucleus, two protons, five gamma-ray photons, and two neutrinos. (See Figure 6.3.)

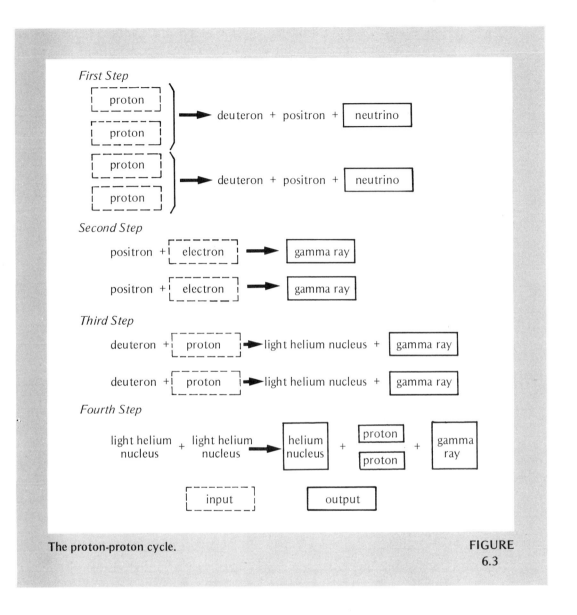

The proton-proton cycle.

FIGURE
6.3

The carbon-nitrogen cycle produces much less of the sun's energy than the proton-proton cycle. It requires temperatures of about 30,000,000 degrees Kelvin to become the dominant reaction, and the interior of the sun is not this hot. The presence of carbon nuclei ($_6C^{12}$) is required too, although the carbon nuclei remain at the completion of the cycle. It begins with the collision of a proton and the nucleus of an ordinary carbon atom, which produces a light isotope of nitrogen and a gamma-ray photon:

$$_1H^1 + {_6}C^{12} \rightarrow {_7}N^{13} + \text{gamma ray}$$

A carbon nucleus experiences such a collision once in about 2,000,000 years, on the average.

The light nitrogen is an unstable isotope and quickly disintegrates to form a heavy isotope of carbon, a positron, and a neutrino:

$$_7N^{13} \rightarrow {_6}C^{13} + \text{positron} + \text{neutrino}$$

The neutrino escapes into space. The positron quickly collides with an electron; they annihilate one another to produce a gamma-ray photon:

$$\text{positron} + \text{electron} \rightarrow \text{gamma ray}$$

Within minutes, the nucleus of the heavy carbon isotope collides with a proton. This produces the nucleus of an ordinary nitrogen atom and a gamma-ray photon:

$$_6C^{13} + {_1}H^1 \rightarrow {_7}N^{14} + \text{gamma ray}$$

The average life of such a nitrogen nucleus is about 4,000,000 years. It eventually collides with a proton to produce the nucleus of a light oxygen isotope and a gamma-ray photon:

$$_7N^{14} + {_1}H^1 \rightarrow {_8}O^{15} + \text{gamma ray}$$

The oxygen isotope is unstable. It disintegrates within minutes to produce the nucleus of a heavy nitrogen isotope, a neutrino, and a positron:

$$_8O^{15} \rightarrow {_7}N^{15} + \text{positron} + \text{neutrino}$$

Again, the neutrino escapes into space, and the positron is quickly annihilated in a collision with an electron, leaving a gamma-ray photon:

$$\text{positron} + \text{electron} \rightarrow \text{gamma ray}$$

Within a few years, the heavy nitrogen isotope collides with a proton. The products of the reaction are nuclei of ordinary helium and ordinary carbon atoms and a gamma-ray photon:

$$_7N^{15} + {_1}H^1 \rightarrow {_2}He^4 + {_6}C^{12} + \text{gamma ray}$$

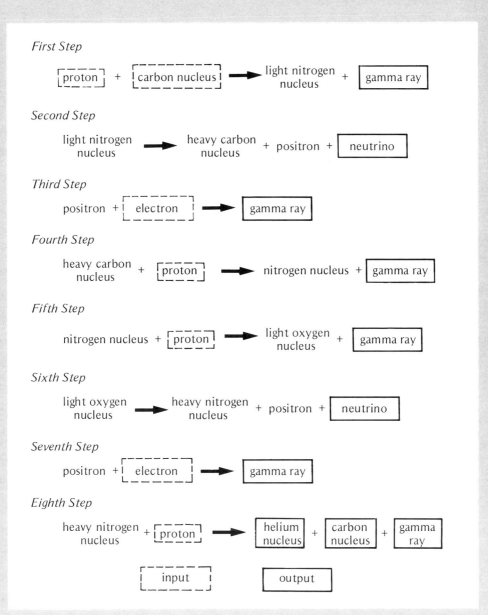

First Step

proton + carbon nucleus ➝ light nitrogen nucleus + gamma ray

Second Step

light nitrogen nucleus ➝ heavy carbon nucleus + positron + neutrino

Third Step

positron + electron ➝ gamma ray

Fourth Step

heavy carbon nucleus + proton ➝ nitrogen nucleus + gamma ray

Fifth Step

nitrogen nucleus + proton ➝ light oxygen nucleus + gamma ray

Sixth Step

light oxygen nucleus ➝ heavy nitrogen nucleus + positron + neutrino

Seventh Step

positron + electron ➝ gamma ray

Eighth Step

heavy nitrogen nucleus + proton ➝ helium nucleus + carbon nucleus + gamma ray

input   output

The carbon-nitrogen cycle.

FIGURE
6.4

The complete cycle, which results from the interaction of four protons, one carbon nucleus, and two electrons, produces one helium nucleus, one carbon nucleus, two neutrinos, and six gamma-ray photons. (See Figure 6.4.)

The time required to produce one helium nucleus, with concomitant mass-energy conversion, averages a few billion years in the proton-proton cycle and a few million years in the carbon-nitrogen cycle. However, the number of nuclei and particles available for these reactions in the sun is enormous. The sun is losing mass at the rate of more than 4,000,000 tons per second, and still it has a several-billion-year supply of hydrogen.

HERTZSPRUNG-RUSSELL DIAGRAM

Pursuit of questions about the birth, life, and death of stars requires identification and analysis of patterns in observational data. Astronomers have found it helpful to analyze the pattern revealed by plotting temperature against luminosity for a large number of stars. In 1911 and 1913, respectively, Ejnar Hertzsprung and Henry Norris Russell made such plots. Hertzsprung plotted color-magnitude data for stars in the Hyades cluster; Russell plotted magnitude-spectral class data for stars of known parallax. Such displays of data are now referred to as Hertzsprung-Russell or H-R diagrams. H-R digrams will greatly facilitate our subsequent discussion of stellar evolution.

Figure 6.5 is an H-R diagram. Rather than plotting the points from enough individual stars to reveal the pattern, we have simply indicated the general areas where most of the points would be grouped if plotted. When the data for all of the stars within several parsecs of the sun are plotted, the points cluster in four particular portions of the diagram. By far the largest number of stars lie along a path extending from the upper left to the lower right. This portion of the H-R diagram is referred to as the *main sequence.* The sun is a main-sequence star. Smaller numbers of stars are grouped in three other areas of the diagram. White dwarfs—small white-hot but faint stars—are grouped below the main sequence. Giants and supergiants—two groups of cooler (red to yellowish) large stars—occur above the main sequence. The supergiants are rare; Betelgeuse in Orion is an example of a supergiant.

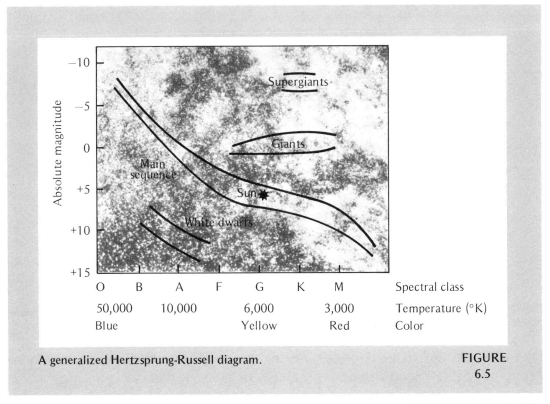

A generalized Hertzsprung-Russell diagram.

FIGURE
6.5

STELLAR LIFE SPANS

A star's hydrogen fuel supply is finite. The sun's mass is about 2×10^{33} g, and it is radiating energy at the rate of about 4×10^{33} ergs/sec. It is losing mass at the rate of about 4×10^{12} g/sec in processes that reduce the mass of the hydrogen by a little less than 1%. Approximately 10% of the inner region of the sun is believed to be hot enough to sustain nuclear fusion cycles. All in all, it appears that the sun can radiate in its present manner for perhaps another 5 to 10 billion years. Radiological dating suggests, as we discussed in Chapter 3, that the solar system is nearly 5 billion years old. So the sun seems to be a middle-aged star.

Is the sun typical? Do all stars burn hydrogen for 10 to 15 billion years? No, indeed. Consideration of masses and luminosities of stars indicates that they consume their nuclear fuel in life spans ranging from a few million years to several trillion years. The more massive and luminous stars have much shorter life spans than the less massive, cooler stars. The massive stars have more hydrogen, but they consume it at a much greater rate than the cooler stars. The sun has already existed longer than the total life span of some other stars. Accordingly, all of the stars did not come into existence at the same time; stars are believed to be coming into existence continually. *Coming into existence* is probably more suitable terminology than *creation*. *Creation* would connote ultimate explanations of beginnings, which we are not prepared to offer.

Stars are apparently of many different ages and have many different life spans, and yet most stars are on the main sequence in the H-R diagram. This suggests that a star spends most of its life span on the main sequence. Does it follow that stars not on the main sequence are at a different stage in their development than stars on the main sequence? Astronomers believe it does.

Stellar Birth

Hydrogen atoms pervade the universe, so the basic stellar building material is available in space. Stars also contain lesser amounts of other elements. We know that some of these elements are available in space, because we know of several different kinds of interstellar molecules. Space also contains interstellar dust; clouds of interstellar dust can be seen silhouetted against the light from more distant objects. The dust particles are of unknown composition; they are probably composed of the same kinds of atoms that make up the interstellar gas molecules. They are tiny, with diameters comparable to the wavelengths of light. We indicated in Chapter 3 that even the bright and dark nebulae are as rarefied as the best laboratory vacuums. Therefore, one might wonder whether there is enough interstellar gas and dust to form stars, if it could be brought together. There is ample material for the formation of stars. The Great Nebula in Orion alone contains enough material to form more than 1,000 stars as massive as the sun.

Nebulae are the birthplaces of stars; it is in these clouds of dust and gas that the raw materials for stellar formation are most abundant. (We will skirt the question of the source of the nebulae.) Atoms and molecules are in constant, random motion; their kinetic energy is their heat energy content. Chance will sometimes increase the concentration of the randomly moving atoms, molecules, and other particles in certain regions of the nebulae. If the concentration is sufficient, the gravitational attraction among the moving particles becomes significant and prevents their dispersal. Concentrations of gas and dust develop and swirl about within nebulae. If the density of hydrogen and other matter becomes great enough, gravitational forces cause the material to begin to contract. This volume of contracting materials is a *protostar;* the diameter of protostars is typically a sizeable fraction of a light year. Many photographs of nebulae reveal dark *globules,* which astronomers believe to be regions where protostars are forming. (See Figure 6.1.)

As the protostar contracts it heats up, because material falls inward with increasing speed. The kinetic energy of the falling material gives rise to radiation. Eventually the protostar becomes visible as a red object, perhaps a rather bright one. If one located it on the H-R diagram, it would be well toward the right side and above the main sequence. When the density and temperature have mounted within the protostar, a period of somewhat erratic behavior ensues. At times the inward gravitational forces dominate, and at other times expansive heat forces dominate. A class of unpredictable variable stars called the T Tauri stars are thought to be in this stage of development. In general, new stars move leftward or perhaps downward and leftward in the H-R diagram toward the main sequence. Leftward movement indicates increasing temperature. This means that heat energy is developing faster in the contracting protostar than it is leaving as radiation. Eventually, temperatures at the core reach the point at which the proton-proton cycle can be sustained. Expansive forces develop at the core through nuclear fusion and reach an equilibrium with the inward gravitational forces (the weight) of the cooler gases in the outer regions. Contraction ceases. The star is now on the main sequence. It spends most of its life on the main sequence, burning hydrogen at a fairly steady rate. Helmholtz's old theory has still proven useful in explaining how the star's nuclear furnace was ignited.

The speed with which a contracting protostar moves to the main sequence depends upon its mass. More massive protostars contract more rapidly, because greater gravitational forces develop within them. Also, more massive stars "arrive" at a higher point on the main sequence than less massive stars; their temperatures and luminosities are higher when they become main-sequence stars. Their temperatures may be great enough to sustain the swifter carbon-nitrogen cycle. We have indicated that more massive stars consume their hydrogen more quickly. Massive protostars contract to main-sequence size in less than 100,000 years;

the sun probably contracted for more than 50,000,000 years; stars with relatively little mass contract for billions of years. Figure 6.6 shows in a very general way the tracks that would be formed on the H-R diagram if one plotted the positions of protostars of different masses throughout their period of contraction.

FIGURE
6.6

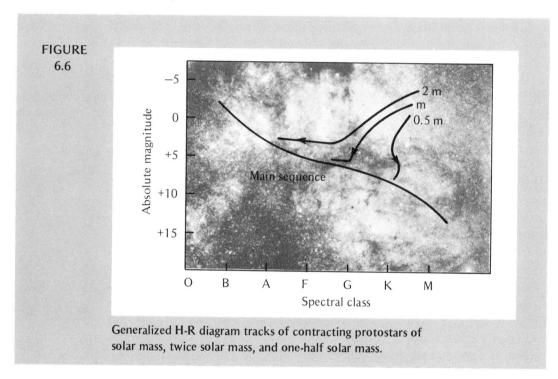

Generalized H-R diagram tracks of contracting protostars of solar mass, twice solar mass, and one-half solar mass.

Stellar Death

What happens to a main sequence star when it runs out of hydrogen? Does it go out quietly, like a candle that has consumed its wax? No, before fading away, it puts on a dramatic display. It may literally go out with a bang.

The stars in *galactic (open) clusters,* such as the Pleiades and Hyades, are believed to have come into existence at essentially the same time. As described in Chapter 3, they move through space together. Further suggestive of their common origin is the fact that they are often associated with nebular matter, which could be the substance from which they were born. Even though the protostars for cluster members came into existence at about the same time, stars in the clusters have different life spans and therefore are in different stages of their development. This conclusion follows from the fact that stars in the cluster differ in mass. Astronomers reasoned that H-R diagrams of stars in galactic clusters should reveal something of the behavior of aging stars.

H-R diagrams of galactic clusters are indeed revealing. The stars farthest along in their life spans (those that reached the main sequence first) have frequently left the main sequence. They have moved upward, or upward and to the right. The older the cluster, the greater the proportion of stars that have left the main sequence. Figure 6.7 indicates the general appearance of H-R diagrams for open clusters of increasing age. Movement upward indicates brightening. Movement to the right indicates cooling and reddening. A star that has moved far enough upward from and to the right of the main sequence has brightened and reddened enough to become a red giant or a supergiant. How?

Hydrogen fusion is confined largely to the central core of the star during the period it spends on the main sequence. Energy moves outward from the core by radiation rather than convection, so there is little mixing of stellar material. The composition of the core changes from largely hydrogen to largely helium. As the end of the core's hydrogen supply approaches, the fusion processes slow and the gravitational forces within the core become greater than the outward expansive forces. The core begins to contract. Even though nuclear fusion has stopped in the core, the temperature of the core rises as it contracts. Heat moving outward from the core raises the temperature of unburned hydrogen surrounding the core. Hydrogen-to-helium fusion then begins to spread outward from the core. As this activity moves outward, the star brightens and expands. The star moves away from the main sequence, and it swells until its diameter is many times what it was while on the main sequence. The total outpouring of energy becomes much greater, as the increased luminosity indicates, but the surface area increases greatly too. Since the radiation is leaving the star from vastly more square units of surface area, the radiation per unit of surface area is less and the star is redder. It becomes a red giant or a supergiant. A star of the sun's mass might remain a giant for 100,000,000 years. What happens next?

Astronomers are less confident that they know what happens next, but they have some hypotheses. Down in the core the helium has

FIGURE
6.7

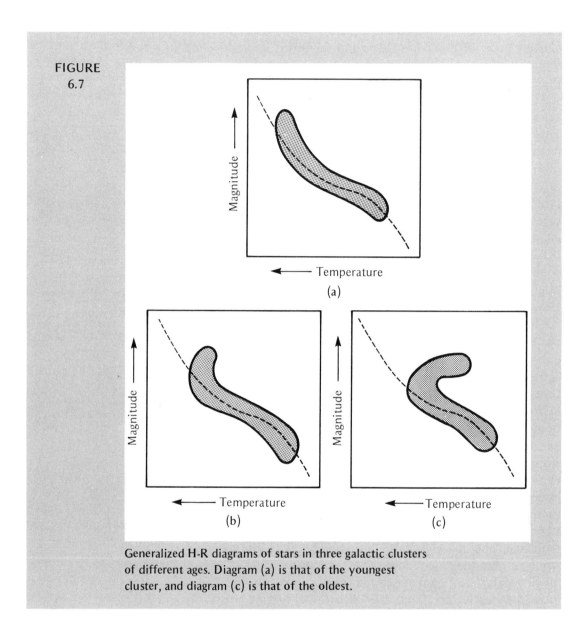

Generalized H-R diagrams of stars in three galactic clusters
of different ages. Diagram (a) is that of the youngest
cluster, and diagram (c) is that of the oldest.

contracted under gravitational forces, and the temperature of the core
has risen dramatically to 100,000,000 degrees Kelvin. This is hot
enough to sustain helium burning, another nuclear fusion cycle. Two
helium nuclei fuse to form an unstable beryllium isotope. Many of the

beryllium nuclei promptly disintegrate, releasing the helium nuclei. But sometimes, before it can disintegrate, a beryllium nucleus combines with another helium nucleus to produce a carbon nucleus and two gamma-ray photons. A carbon nucleus may fuse with a helium nucleus to produce an oxygen nucleus, and the oxygen nucleus may combine with another helium nucleus to produce a neon nucleus. Gamma rays are also released in these reactions. Thus, helium burning not only releases energy, it also synthesizes some heavier elements. These new elements can fuse in still other cycles, releasing energy and synthesizing still heavier elements. When helium burning starts, the star begins to move leftward on the H-R diagram, away from its red giant location and toward the main sequence. Leftward movement means that the star is becoming hotter and bluer in color.

As the star's position moves leftward on the H-R diagram, it eventually crosses the main sequence, but at a point higher up on the sequence than the point from which it departed long before. For a star of the sun's mass, this period of evolution from red giant to old blue star might require about 100,000 years. By the time the star's position has crossed the main sequence moving leftward, it is about out of nuclear fuel. If its mass is not much more than that of the sun, the star begins to fade away. Its position drops down the left side of the H-R diagram and moves somewhat to the right. The star loses luminosity and becomes a white dwarf. Nuclear fusion has ceased. The star continues to emit some light during a very long cooling-off period. It may eventually become an unseen "black dwarf." Our Milky Way Galaxy may be too young to contain any black dwarfs yet.

Some aging stars are unable to drop quietly into white-dwarf status; they are too massive to do so. Subrahmanyan Chandrasekhar, an Indian astrophysicist working at Yerkes Observatory, theorized that the more massive a white dwarf, the smaller its diameter. Strangely, then, a dwarf of very great mass might have essentially no diameter. Chandrasekhar calculated the maximum masses of stellar matter within which pressures could develop to resist such theoretically total gravitational collapse. The *Chandrasekhar limit* varies with the kinds of atoms present in the star; for stars composed of relatively heavy elements, it is about 1.4 solar masses. A star more massive than this must lose mass if it is to become a white dwarf. It can lose mass in various ways: it might whirl some matter off as it rotates rapidly; it might eject a shell of gas to form a *ring (planetary) nebula;* or it might become a supernova in a final, chaotic outburst of energy from various fusion reactions.

221

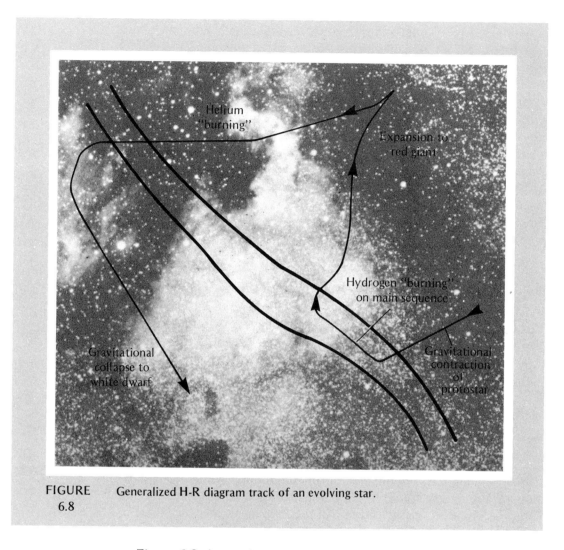

FIGURE   Generalized H-R diagram track of an evolving star.
6.8

Figure 6.8 shows, in a very general way, the changing H-R diagram positions of an evolving star of moderate mass.

NEUTRON STARS AND BLACK HOLES

White-dwarf status is thought to be the evolutionary fate of most but not all stars. Decaying stars within certain ranges of mass are believed to develop internal forces that cause them to contract until they have densities greater than those of white dwarfs.

Some scientists postulate the existence of black holes, or stars that result from ultimate contraction of the matter of which they are constituted. A black hole with a mass equal to that of the sun would be less than ten kilometers in diameter and have a density greater than 10^{16} g/cm^3. The gravitational field of a black hole would be so intense that no light or other radiation could escape from it. Not all astronomers believe that black holes can exist. Some believe, however, that a few black holes have already been detected by their influence on x-ray sources in their vicinity.

Another possible fate for an aging star is becoming a neutron star. Neutron stars are believed to be formed by supernovae. It is thought that tremendous pressures force together the protons and electrons within atomic nuclei to form neutrons. Astrophysicists believe that rapidly rotating neutron stars are being observed as pulsars. (These stars were discussed in more detail in Chapter 3.)

STELLAR POPULATIONS I AND II

Globular clusters can be seen to form a halo incasing the disks of several nearby galaxies, as described in Chapter 3. Baade observed that stars in the spiral arms of the Andromeda galaxy are predominantly blue and blue-white, whereas those in the center of the galaxy and in the halo are from white to red. He categorized the stars in the spiral arms as Population I stars and those in the center and the halo as Population II stars. This useful generalization is applied to stars in our Milky Way Galaxy and other galaxies. The concept of stellar populations has been refined and extended, and astronomers sometimes utilize more than these original two categories. Population I stars occur in the midst of gas and dust, the stellar building materials; they are believed to be relatively young stars. Population II stars occur in "cleaner" regions, where there is little gas and dust. They are believed to be older stars.

GALACTIC EVOLUTION

How did our Milky Way Galaxy evolve? Perhaps its evolution began somewhat like that of a star, but on a bigger scale. The gas cloud from which the *protogalaxy* condensed experienced gravitational contraction, like a protostar. Within the contracting protogalaxy, eddies of gas

FIGURE
6.9

Barred spiral galaxy in Eridanus: NGC 1300. Photograph
was taken with the 200-inch telescope. (Photograph courtesy of the Hale Observatories.)

developed and swirled about. In some places the concentration of gas
became great enough for the process of stellar evolution to begin.
Sometimes an accumulation of protostar material within the contracting protogalaxy was sufficiently massive to spawn a great cluster of
stars. Many of the first stars to come into existence were those that we
now see out in the halo of globular clusters; these developed about 10
billion years ago. The protogalaxy was rotating; as it contracted, it
rotated faster. The increase in the speed of rotation flattened the gas
and dust into a disk. Stars are still being formed around us in the gas
and dust that remain in the spiral arms of the flattened disk.

Irregular, spiral, and elliptical galaxies were mentioned in Chapter 3.
Astronomers have developed a more detailed classification scheme,
which recognizes several shapes of spiral and elliptical galaxies. Among
the spirals are several forms of *barred spirals.* Figure 6.9 is a picture of a
barred spiral.

Irregular galaxies are usually rich in gas, dust, and young-looking
blue stars. Elliptical galaxies typically consist of red stars and practi-
cally no gas and dust. Therefore, it is tempting to conclude that galaxies
evolve from a youthful, irregular stage through a middle-aged, spiral
stage to an old, elliptical stage. But there are too many exceptions to
such a neat conclusion. Some irregular galaxies contain red stars and

little gas, and most galaxies seem to contain some old stars. This is interpreted by some to indicate that all galaxies came into existence at about the same time, but developed differently. We have considered how stars of different masses can come into existence at the same time and age at different rates. Perhaps protogalaxies that came into existence at the same time but had different masses and rotational speeds produced galaxies of different shapes within which stars are developing at different rates. Astronomers are confronted by numerous unanswered questions about galactic evolution.

CONCLUSION

We began this discussion of astronomy by observing the general appearance of the sky and by observing that we seem to be motionless at the center of a turning celestial sphere. From a searching consideration of the appearance of things, we learned a great deal. And this discussion has described only a small part of what astronomers have learned.

We find that we are not at the center of the universe. We are on a small ball of rock that orbits an ordinary star among 100 billion stars in a galaxy that is but one of billions of galaxies. The little earth on which we ride has apparently been orbiting the sun for nearly 5 billion years, and it appears destined to do so for a few billion more, until the sun swells and reddens and incinerates all living things on the earth. Our bodies are literally the dust of the earth, brought together for a while in an incredibly complex way. Perhaps the atoms in our bodies were synthesized in the nuclear furnace of an ancient star, hurled into the void by a supernova, and eventually included in the cloud of gas from which the sun and earth condensed. The mysterious spark of life, which we can transmit but not ignite, is sustained in us by the radiant energy of the star we orbit. We live for a moment in the eons of time, which stretch into the past and future; then the atoms in our bodies return to the earth and to the store of cosmic building material.

So much remains to be learned! We still don't even know whether or not we are alone in the universe. When we look out into space, are intelligent creatures looking back? If not, is there any kind of living thing existing anywhere other than on earth? Most of us hope so, and most astronomers think so. Could this just be wishful thinking? It is more exciting to speculate about life in other planetary systems than to

FIGURE
6.10

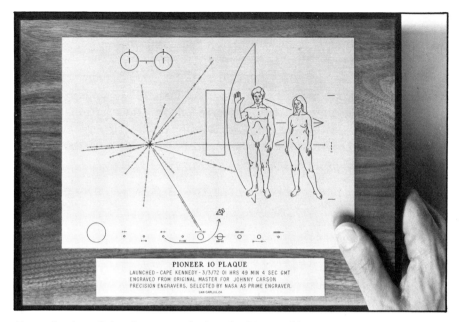

Reproduction of the plaque attached to the Pioneer 10 spacecraft. It is designed to show inhabitants of other star systems when Pioneer was launched, from where, and by what kind of beings. The radiating lines indicate the directions of 14 pulsars relative to the sun, the home star of the launching civilization. Binary numbers on the radiating lines indicate the frequencies of the pulsars at the time of launch relative to that of the hydrogen atom, which is symbolized above them. The binary number (8) next to the woman, multiplied by the hydrogen wavelength (21 cm) gives the height of the woman. (Photograph courtesy of Precision Engravers, San Carlos, California.)

believe that we are alone in the vast universe. But science is based on the conviction that nature is not capricious, and therefore the processes that led to life here should operate under similar conditions anywhere in the universe. Statistically, earth-like conditions should exist in countless other places. Will we hear from living things on other celestial objects? We have made a few serious efforts to listen for radio signals from intelligent creatures in other planetary systems, and there will be more attempts. Will we be able to travel to other planetary systems? At this time, there seems to be an insurmountable barrier prohibiting travel beyond our solar system. This barrier is the concept of the speed of light as a limiting velocity. It seems that as an object's speed increases, so does its inertia; inertia apparently becomes infinite at the speed of light. The distances involved in travel outside the solar system require speeds approaching that of light; the attainment of such speeds would require enormous supplies of energy. Of course, barriers that seemed insurmountable in the past have been scaled. Is our first message to extra-terrestrial creatures contained in the plaque attached to the Pioneer 10 unmanned spacecraft? It was hurled from the earth on March 3, 1972 with sufficient velocity to escape from the solar system. (See Figure 6.10.)

Will the galaxies eventually cease their outward rushing? Will gravitation cause them to fall inward into a collapsing universe? Will we eventually be unable to see farther, even with larger telescopes? Have we already seen to the edge of the universe?

Is space infinite or finite? If it is finite, what is its extent and structure? Is the universe dying? Mass is being converted to energy in processes of nuclear fusion; will the mass be replaced? The second law of thermodynamics indicates that energy transformations ultimately result in heat (the kinetic energy of randomly moving particles of matter), and heat flows from hotter to cooler bodies. Will the entire universe eventually be dark and cold? Does the second law of thermodynamics decree the ultimate fate of the universe? How did it begin? We believe that we know a considerable amount about our place in the universe, but what is our purpose in it?

Many are humbled by the contemplation of unanswered questions. Perhaps their feeling is akin to that which Job experienced when God thundered at him from out of the whirlwind:

Where were you when I laid
   the foundations of the earth?
Tell me, if you have understanding. . . .

Have you commanded the morning
   since your days began,
and caused the dawn to know its place? . . .

Where is the way to the dwelling
   of light,
and where is the place of darkness? . . .

Can you bind the chains
   of the Pleiades,
or loose the cords of Orion? . . .

Do you know the ordinances
   of the heavens?

FOR REVIEW, DISCUSSION, OR FURTHER STUDY

1. Describe the dominant process by which the sun releases energy.

2. Describe the evolution of a star whose mass is about equal to that of the sun.

3. In what stage of stellar evolution are T Tauri stars believed to be?

4. What has study of H-R diagrams of galactic clusters suggested about stellar evolution?

5. Why are Population II stars believed to be older than Population I stars?

6. Can it be concluded logically that spiral galaxies are older than irregular galaxies and that elliptical galaxies are older than spiral galaxies?

7. More than 60 of the elements found on the earth have been detected on the sun. Were they formed on the sun or did they come from elsewhere?

8. What is the apparently insurmountable barrier that presently seems to essentially eliminate the possibility of travel to planetary systems beyond our solar system?

9. What are Carl Sagan's views on whether or not the earth has been visited by spacecraft from other planets?

10. Is there a limit to how far we can expect to see, regardless of the size of the telescope?

SUGGESTED ACTIVITIES

What Is the Shape of Space?

Cut a one-inch strip of paper from the long edge of a sheet of typing paper. Mark the midpoint of one side of the strip with an X and the midpoint of the other side with an O. Bring the ends of the strip together to form a squatty cylinder or "bracelet." Observe that there is no way in which a path can be traced from the X through the O and back to the X without passing around an edge of the strip; the cylinder clearly has outside and inside surfaces. Straighten the strip. Bring the ends together again, but before joining them give one end of the strip a half twist (180 degrees). Now trace a path from X through O and back to X. Note that it is possible to trace the path without passing around an edge of the strip; this structure does not have inside and outside surfaces. This demonstration is not intended to suggest that space is shaped like this Möbius strip. Rather, the point is to suggest that those of us who find it difficult to contemplate the extent and shape of space should be mindful that unconventional geometry may be involved in these cosmological matters.

SUGGESTED READINGS

Alfven, Hannes, "Plasma Physics, Space Research, and the Origin of the Solar System," *Science*, Vol. 172 (4 June 1971), 991.

Arp, Halton, "Observational Paradoxes in Extragalactic Astronomy," *Science*, Vol. 174 (17 December 1971), 1189.

Gamow, George, "The Evolutionary Universe," *Scientific American*, Vol. 195, No. 3 (September 1956), 136.

Gott, J. Richard, III, James E. Gunn, David N. Schramm, and Beatrice M. Tinsley, "Will the Universe Expand Forever?" *Scientific American*, Vol. 234, No. 3 (March 1976), 80.

Greenstein, Jesse L., "Dying Stars," *Scientific American*, Vol. 200, No. 1 (January 1959), 46.

Hammond, Allen L., "Stellar Old Age: White Dwarfs, Neutron Stars, and Black Holes," *Science*, Vol. 171 (12 March 1971), 994.

Hammond, Allen L., "Stellar Old Age (II): Neutron Stars and Pulsars," *Science*, Vol. 171 (19 March 1971), 1133.

Hammond, Allen L., "Stellar Old Age (III): Black Holes and Gravitational Collapse," *Science*, Vol. 171 (26 March 1971), 1228.

Huang, Su-Shu, "Life Outside the Solar System," *Scientific American*, Vol. 202, No. 4 (April 1960), 55.

Hynek, J. Allen, Letter to *Science*, Vol. 154 (21 October 1966), 329 (in which he suggests that UFO's merit scientific study).

Lewis, John S., "The Chemistry of the Solar System," *Scientific American*, Vol. 230, No. 3 (March 1974), 50.

Metz, William D., "The Decline of the Hubble Constant: A New Age for the Universe," *Science*, Vol. 178 (10 November 1972), 600.

Pasachoff, Jay M., and William A. Fowler, "Deuterium in the Universe," *Scientific American*, Vol. 230, No. 5 (May 1974), 108.

Peebles, P. J. E., and David T. Wilkinson, "The Primeval Fireball," *Scientific American*, Vol. 222, No. 6 (June 1970), 26.

Rees, Martin, and Joseph Silk, "The Origin of Galaxies," *Scientific American*, Vol. 222, No. 6 (June 1970), 26.

Sagan, Carl. *The Cosmic Connection*. New York: Dell Publishing Co., Inc., 1973, 267 pages.

Sagan, Carl, "The Past and Future of American Astronomy," *Physics Today*, Vol. 27, No. 2 (December 1974), 23.

Sagan, Carl, and Frank Drake, "The Search for Extraterrestrial Intelligence," *Scientific American*, Vol. 232, No. 5 (May 1975), 80.

Sagan, Carl, and Thornton Page (eds.). *UFO's—A Scientific Debate.* Ithaca, New York: Cornell University Press, 1972, 310 pages.

This volume contains papers presented at a symposium on Unidentified Flying Objects sponsored by the American Association for the Advancement of Science.

Strom, Richard G., George K. Miley, and Jan Oort, "Giant Radio Galaxies," *Scientific American,* Vol. 233, No. 2 (August 1975), 26.

Thomsen, Dietrick E., "Black Holes: No Longer Hypothetical," *Science News,* Vol. 103 (13 January 1973), 28.

Thomsen, Dietrick E., "The Blob That Ate Physics," *Science News,* Vol. 108 (12 July 1975), 28.

Thomsen, Dietrick E., "Classifying Radio Galaxies," *Science News,* Vol. 108 (21 September 1975), 204.

Thomsen, Dietrick E., "Cosmology According to Hoyle," *Science News,* Vol. 107 (14 June 1975), 386.

Thomsen, Dietrick E., "The Weak Interaction in the Universe," *Science News,* Vol. 107 (31 May 1975), 354.

Thorne, Kip S., "The Search for Black Holes," *Scientific American,* Vol. 231, No. 6 (December 1974), 32.

Ulrich, Roger K., "Solar Neutrinos and Variations in the Solar Luminosity," *Science,* Vol. 190 (14 November 1975), 619.

Wick, Gerald L., "Neutrino Astronomy: Probing the Sun's Interior," *Science,* Vol. 173 (10 September 1971), 1011.

Appendixes

Appendix A STAR CHARTS

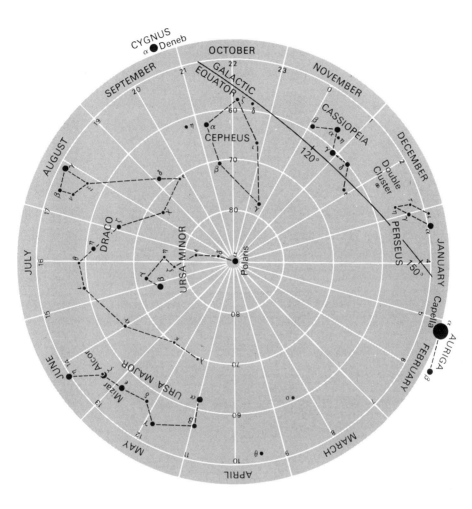

SCALE OF MAGNITUDES

MAP 1. The North Polar Constellations
(From Stanley P. Wyatt, Principles of
Astronomy, *2nd ed., p. 86. Copyright
1971 by Allyn and Bacon, Inc.).*

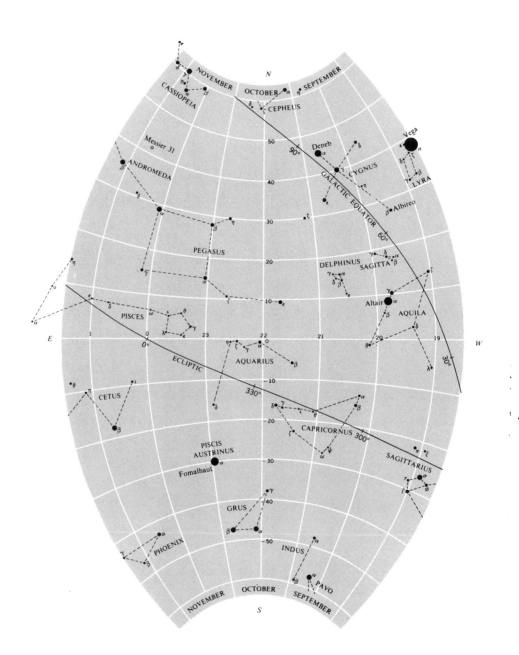

MAP 2. *The Constellations of Autumn (Wyatt, p. 88).*

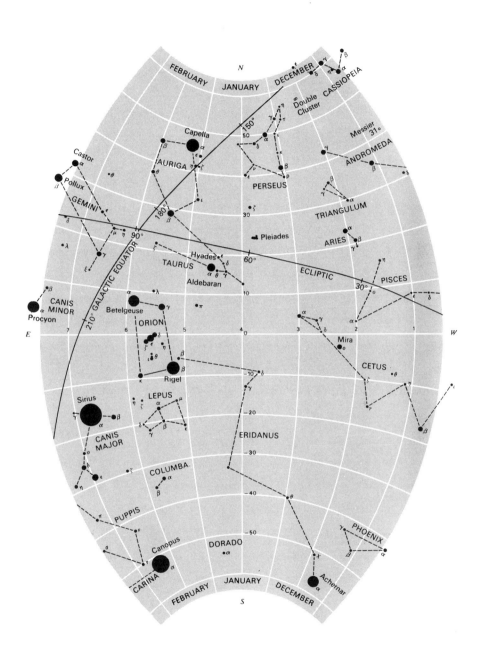

MAP 3.   The Constellations of Winter   (*Wyatt, p. 89*).

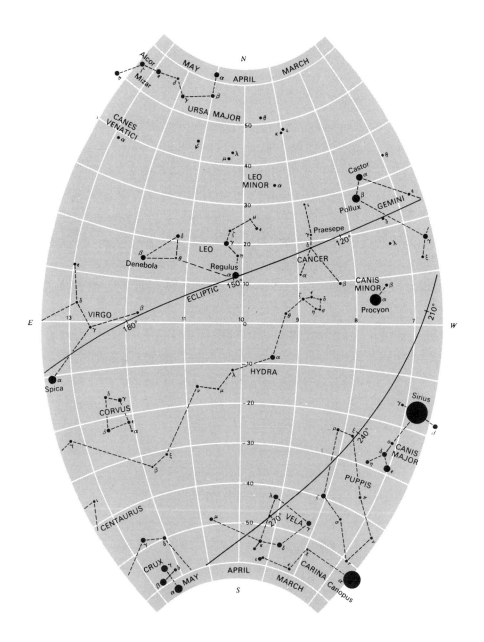

MAP 4. The Constellations of Spring (*Wyatt, p. 90*).

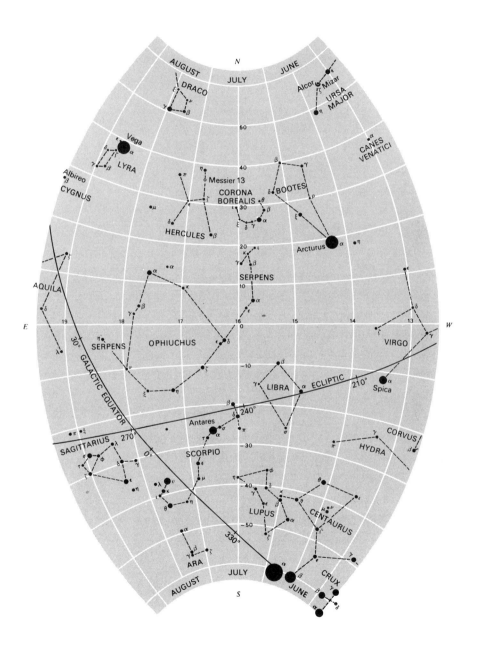

MAP 5. The Constellations of Summer *(Wyatt, p. 91)*.

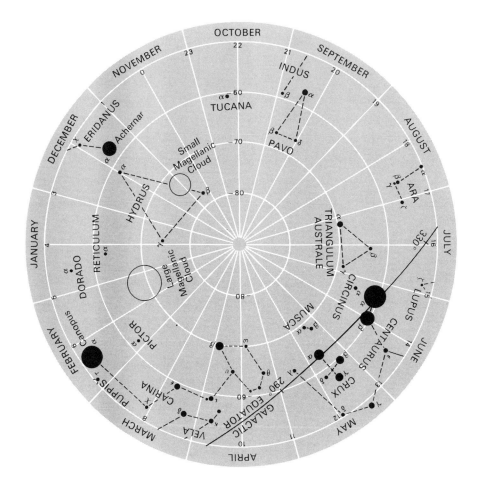

MAP 6. The South Polar Constellations   (*Wyatt, p. 87*).

Appendix B THE BRIGHTEST STARS

The brightest stars visible from mid-continental United States are listed below. Some of these stars vary in brightness and may appear in a slightly different order in other lists. Several of the larger distances are estimates based on spectral characteristics.

Star	Location	Apparent Magnitude	Distance (in light years)
Sirius	in Canis Major	−1.42	8.7
Arcturus	in Boötes	−0.06	36
Vega	in Lyra	0.04	26
Capella	in Auriga	0.05	45
Rigel	in Orion	0.14	900
Procyon	in Canis Minor	0.37	11
Betelgeuse	in Orion	0.41	520
Altair	in Aquila	0.77	16
Aldebaran	in Taurus	0.86	67
Spica	in Virgo	0.91	250
Antares	in Scorpius	0.92	380
Pollux	in Gemini	1.16	35
Fomalhaut	in Piscis Austrinus	1.19	22
Deneb	in Cygnus	1.26	1400
Regulus	in Leo	1.36	83

Appendix C CONSTANTS AND CONVERSION FACTORS

1 mile = 1 mi = 5280 feet = 1.6093 kilometers
1 kilometer = 1 km = 1000 meters = 0.6214 mile
1 meter = 1 m = 100 centimeters
1 centimeter = 1 cm = 10 millimeters
1 millimeter = 1 mm = 1000 microns
1 gram = 1 g = 1000 milligrams
1 kilogram = 1 kg = 1000 grams = 2.2046 pounds
1 angstrom = 10^{-10} meter
1 pound = 0.4536 kilogram
speed of light (in vacuum) = 2.998×10^{10} cm/sec
$ = 186,300$ mi/sec
1 light year = 9.460×10^{17} cm = 5.878×10^{12} mi
1 parsec = 3.262 light years
1 astronomical unit = 1.496×10^8 km = 92,960,000 mi
degrees Fahrenheit = 1.8 × degrees centigrade + 32
degrees centigrade (Celsius) = $\frac{5}{9}$(degrees Fahrenheit − 32)
degrees Kelvin = degrees centigrade + 273
1 tropical (ordinary) year = 365.2422 mean solar days
mass of the earth = 5.98×10^{27} g
mass of the sun = 1.99×10^{33} g
diameter of the sun = 13.92×10^8 km

Appendix D THE CONSTELLATIONS

Latin Name	English Name	Right Ascension (in hours)	Declination (in degrees)
Andromeda	Princess (of Ethiopia)	1	+40
Antlia	Air Pump	10	−35
Apus	Bird of Paradise	16	−75
Aquarius	Water Bearer	23	−15
Aquila	Eagle	20	+5
Ara	Altar	17	−55
Aries	Ram	3	+20
Auriga	Charioteer	6	+40
Boötes	Herdsman	15	+30
Caelum	Graving Tool	5	−40
Camelopardalis	Giraffe	6	−70
Cancer	Crab	9	+20
Canes Venatici	Hunting Dogs	13	+40
Canis Major	Big Dog	7	−20
Canis Minor	Little Dog	8	+5
Capricornus	Goat	21	−20
Carina	Keel (of Argo)	9	−60
Cassiopeia	Queen (of Ethiopia)	1	+60
Centaurus	Centaur	13	−50
Cepheus	King (of Ethiopia)	22	+70
Cetus	Whale	2	−10
Chamaeleon	Chameleon	11	−80
Circinus	Compasses	15	−60
Columba	Dove	6	−35
Coma Berenices	Bernice's Hair	13	+20
Corona Australis	Southern Crown	19	−40
Corona Borealis	Northern Crown	16	+30
Corvus	Crow	12	−20
Crater	Cup	11	−15
Crux	Cross (Southern)	12	−60
Cygnus	Swan	21	+40
Delphinus	Dolphin	21	+10

Latin Name	English Name	Position	
		Right Ascension (in hours)	Declination (in degrees)
Dorado	Swordfish	5	−65
Draco	Dragon	17	+65
Equuleus	Little Horse	21	+10
Eridanus	River	3	−20
Fornax	Furnace	3	−30
Gemini	Twins	7	+20
Grus	Crane	22	−45
Hercules	Hercules	17	+30
Horologium	Clock	3	−60
Hydra	Water Serpent	10	−20
Hydrus	Water Snake	2	−75
Indus	Indian	21	−55
Lacerta	Lizard	22	+45
Leo	Lion	11	+15
Leo Minor	Little Lion	10	+35
Lepus	Hare	6	−20
Libra	Balance Scale	15	−15
Lupus	Wolf	15	−45
Lynx	Lynx	8	+45
Lyra	Lyre (Harp)	19	+40
Mensa	Table Mountain	5	−80
Microscopium	Microscope	21	−35
Monoceros	Unicorn	7	−5
Musca	Fly	12	−70
Norma	Level (or Ruler)	16	−50
Octans	Octant	22	−85
Ophiuchus	Serpent Carrier	17	0
Orion	Orion (the Hunter)	5	+5
Pavo	Peacock	20	−65
Pegasus	Pegasus (the Winged Horse)	22	+20
Perseus	Perseus	3	+45
Phoenix	Phoenix	1	−50
Pictor	Easel	6	−55
Pisces	Fishes	1	+15
Piscis Austrinus	Southern Fish	22	−30
Puppis	Stern (of Argo)	8	−40

Latin Name	English Name	Position	
		Right Ascension (in hours)	Declination (in degrees)
Pyxis	Mariner's Compass (of Argo)	9	−30
Reticulum	Net	4	−60
Sagitta	Arrow	20	+10
Sagittarius	Archer	19	−25
Scorpius	Scorpion	17	−40
Sculptor	Sculptor's Tools	0	−30
Scutum	Shield	19	−10
Serpens	Serpent	17	0
Sextans	Sextant	10	0
Taurus	Bull	4	+15
Telescopium	Telescope	19	−50
Triangulum	Triangle	2	+30
Triangulum Australe	Southern Triangle	16	−65
Tucana	Toucan	0	−65
Ursa Major	Big Bear	11	+50
Ursa Minor	Little Bear	15	+70
Vela	Sail (of Argo)	9	−50
Virgo	Virgin	13	0
Volans	Flying Fish	8	−70
Vulpecula	Fox	20	+25

Glossary

absolute magnitude. The apparent magnitude that a celestial body would have at a distance of 10 parsecs (about 32.6 light years).

absolute zero. The temperature at which molecular motion ceases. It is equal to 0 degrees Kelvin or −273 degrees Celsius.

absorption (or dark-line) spectrum. A continuous spectrum crossed by dark lines. When light from a source of a continuous spectrum that has passed through an intervening gas at lower temperature and pressure is examined with a spectroscope, an absorption spectrum is seen.

acceleration. Rate of change of velocity.

achromatic lens. A lens system free of chromatic aberration.

albedo. The percentage of incident sunlight reflected by a solar-system body.

Almagest. An ancient book by Claudius Ptolemy explaining the motions of celestial bodies according to a theory of a geocentric universe.

amplitude. The height of a wave crest or the depth of a wave trough.

angstrom unit. A unit of length equal to 10^{-8} cm.

angular diameter. The angle, measured at the eye of the observer, subtended by the diameter of an object.

angular momentum. The momentum of a body associated with its motion of revolution or rotation.

annular eclipse. A solar eclipse in which a ring of the sun is visible around the moon, because the moon is too distant from the earth to obscure the sun completely.

aphelion. The farthest point from the sun in a solar orbit.

apparent magnitude. A measure of the brightness of a star or other celestial object as seen from the earth.

apparent solar time. The interval of time since the sun last crossed the observer's meridian.

ascending node. The point where an orbiting body, moving northward, intersects a reference plane (usually that of the ecliptic or celestial equator).

asteroid. A minor planet (the largest known asteroid has a diameter of less than 800 km); a planetoid.

astrology. The attempt to demonstrate a relationship between human affairs and the positions of the sun, moon, and planets in relation to the earth.

astronomical unit. A unit of length equal to the mean distance of the earth from the sun; about 1.496×10^8 kilometers or 93,000,000 miles.

atom. The smallest particle of an element that retains the chemical properties of the element.

aurora. Light produced in the earth's upper atmosphere, particularly near the polar regions, because of the action of charged solar particles. Light displays seen in northern and southern skies are the Aurora Borealis (Northern Lights) and Aurora Australis (Southern Lights), respectively.

autumnal equinox. The point where the sun appears to cross the celestial equator moving southward. Fall begins when the sun appears to reach the autumnal equinox.

barred spiral galaxy. A spiral galaxy in which an arm bends away from each end of a bar-shaped formation extending through the nucleus. The bar is actually the straight portion of the two arms.

"big bang" theory. The proposition that the expansion of the universe began with a primeval explosion.

black hole. A star of such extreme density that its surface gravitational attraction prevents the escape of radiation.

Bode's Law (Bode—Titius relationship). A relationship that generates a sequence of numbers giving the approximate mean distances from the sun to the planets, measured in astronomical units.

bolide (or fireball). A very bright meteor.

bright nebula. An emission nebula.

carbon—nitrogen cycle (or carbon cycle). A series of nuclear reactions, catalyzed by carbon, that result in the fusion of hydrogen atoms to form helium atoms, a loss of mass, and the release of energy. It is the principal mechanism by which some stars emit energy.

celestial equator. The great circle on the celestial sphere that is 90 degrees from the north and south celestial poles; the projection of the earth's equator on the celestial sphere.

celestial poles. Points about which the celestial sphere appears to rotate. Extensions of the earth's axis meet the celestial sphere at the celestial poles.

celestial sphere. The apparent sphere of the sky with the observer's eye at the center. Celestial objects appear to be on the surface of the sphere.

Celsius temperature scale. A thermometer scale based on 100 divisions or degrees between the freezing and boiling points of water; the freezing point of water is designated as 0 degrees C, and the boiling point of water is designated as 100 degrees C.

centripetal acceleration. Acceleration toward the center of curving motion.

Cepheid variable. One of a class of pulsating, yellow, giant or supergiant stars with periods of from about 1 to 50 days.

chromatic aberration. The failure of a lens to bring all colors of light passing through it to a single focal point.

chromosphere. The portion of the solar atmosphere directly above the photosphere.

coma (of a comet). Part of the head of a comet; a diffuse cloud surrounding the nucleus.

comet. A small body of frozen and dusty matter in orbit around the sun.

conjunction. A configuration in which a planet has the same right ascension as the sun, or in which any two celestial bodies have the same right ascension.

constellation. One of 88 areas of the sky into which the celestial sphere is divided, or the configuration of stars within the area.

constructive interference. The mutual reinforcement of two waves that are "in step" to form a single wave of greater amplitude.

continuous spectrum. The uninterrupted, rainbow-like band of colors seen when light from an incandescent solid, liquid, or gas under high pressure is examined with a spectroscope.

Copernican system. A concept of the universe in which solar system objects revolve around the sun and in which most observed celestial motion is actually apparent motion caused by the earth's rotation and revolution.

corona. The outermost portion of the sun's atmosphere.

coronagraph. An instrument for photographing the sun's corona by artificially eclipsing the sun's disk in the image formed by the instrument.

cosmology. The study of the total structure of the universe and its evolution.

cosmos. The universe conceived as an embodiment of order and harmony.

Coulomb's force. The force of attraction that exists between unlike electric charges, as well as the force of repulsion between like charges.

crater. A circular depression in the surface of the moon, the earth, or another planet.

cusps of the moon. The points of the crescent moon.

dark nebula. A cloud of interstellar dust seen as a silhouette against luminous objects beyond it.

declination. Angular distance north or south of the celestial equator measured along an hour circle.

deferent. A circle in the Ptolemaic system along which moves the center of another circle (epicycle) traced by a moving point.

descending node. The point where an orbiting body, moving southward, intersects a reference plane, usually that of the ecliptic or celestial equator.

destructive interference. The meeting of two "out of step" waves, resulting in a reduction of their amplitudes.

differential calculus. A branch of mathematics for dealing with rates of change.

diffraction. The bending of light as it passes the edge of an object or through a small opening.

diffuse nebula. A bright or dark irregularly shaped nebula.

Doppler effect. The apparent change in wavelength of sound or electromagnetic radiation when the source and observer are coming closer together or getting farther apart.

eccentricity. The ratio of the distance between the foci to the length of the major axis (the greatest diameter) of an ellipse.

eclipse. An obstruction of all or some of the light from a body caused by another body passing in front of the luminous body.

eclipsing binary star. A binary system in which the orbital plane is seen edge-on, so that periodically the light from one star is diminished as the other passes in front of it.

ecliptic. The apparent path of the sun on the celestial sphere; the plane of the earth's orbit.

electric field. The region around an electrically charged object within which the charge can be detected.

electromagnetic induction. The production of a current in a wire moving in a magnetic field or in a wire within a changing magnetic field.

electromagnetic radiation. Radiant energy propagated by changing electric and magnetic fields. Depending upon the wavelength, the radiation is designated as radio, infrared, light, ultraviolet, X-, or gamma rays.

electromagnetic spectrum. The total array of electromagnetic waves.

electron. A subatomic particle with a negative electric charge and much less mass than a proton or neutron.

element. A substance that cannot be broken down into simpler substances by chemical means.

ellipse. A closed plane curve formed by the intersection of a nonhorizontal plane and a circular cone. The sum of the distances from two fixed points within the ellipse (the foci) to any point on the ellipse is a constant.

elliptical galaxy. A galaxy of ellipsoidal shape, without spiral arms, and containing little interstellar matter.

emission (bright) nebula. A luminous cloud of gas that is excited to fluorescence by the ultraviolet energy from a nearby star.

emission (or bright line) spectrum. A series of bright lines seen when light from an incandescent low-pressure gas is examined with a spectroscope.

energy level. One of the possible energy values that an electron within an atom can possess.

epicycle. A circular orbit of a planet in the Ptolemaic system, the center of which moves on the circumference of a larger circle (the deferent).

equant. A point in the Ptolemaic system that is off-center within a circle and around which either an object or the center of another circle moves.

equation of time. The difference between apparent solar time and mean solar time.

247

equinox. Either of two points of intersection of the ecliptic and the celestial equator.

erg. A unit of energy; equal to the work done by a force of one dyne (10^{-5} newtons) acting through a distance of one centimeter.

faculae. Bright regions on the sun, best seen near the limb.

faulting. Movement of rock along a fracture.

fireball (or bolide). A very bright meteor.

flare. A sudden brightening of an area on the sun due to a tremendous outburst of energy.

focal length. The distance from the center of a lens or a mirror to the focal point.

focal point. The point at which light converged by a lens or mirror comes to a focus.

force. That which can overcome inertia and change the state of motion of a body.

frequency. The number of waves that pass a given point in a unit of time.

galactic (or open) cluster. A loose cluster of stars, numbering from dozens to a few thousand, in the arms or disk of the Milky Way Galaxy.

general theory of relativity. A theory formulated by Albert Einstein dealing with problems that arise when a frame of reference is accelerated with respect to another. It is particularly useful in considering masses moving at speeds approaching that of light.

geocentric. Centered about the earth.

globular cluster. A spherical cluster of many thousands of stars. Such clusters are located in the halos that surround the Milky Way Galaxy and other galaxies.

globule. A small, dense, dark nebula; possibly a protostar.

gnomon. An object of which the length or position of its shadow serves as an indicator.

granules. Small areas of the sun's photosphere, believed to be caused by rising hot gases. They have caused the telescopic appearance of the photosphere to be likened to rice grains.

gravitation. The universal force of attraction between all masses.

gravitational acceleration. The amount of acceleration caused by the gravitational attraction on an object near the surface of a celestial body.

gravity. The gravitational attraction of a celestial body for objects on or near its surface; weight is a measure of gravity.

great circle. A circle that divides a sphere into two equal parts.

Gregorian calendar. A reformed Julian calendar introduced by Pope Gregory XIII in 1582. It provided that the ten days of October 5–14 be dropped from the year 1582, and that centenary years (1800, 1900, etc.) be leap years only when exactly divisible by 400.

harvest moon. The full moon that occurs closest to the time when the sun is at the autumnal equinox.

heliocentric. Centered about the sun.

Hertzsprung-Russell (H-R) diagram. A plot of stellar absolute magnitudes against temperature or spectral class.

horoscope. An astrological chart indicating the positions of the sun, moon, and planets in relation to the place and time of birth of an individual. Astrologers profess that the horoscope has significance for the individual and attempt to assess this significance.

hour circle. A great circle on the celestial sphere passing through the celestial pole.

hunter's moon. The full moon following the harvest moon.

Index of Prohibited Books. A list of books or book passages that the Roman Catholic Church considers dangerous to faith or morals.

index of refraction. A measure of the amount of refraction when light enters a transparent substance obliquely; the ratio of the speed of light in a vacuum to the speed of light in the substance.

inertia. The property of matter that causes it to

resist change in its state of motion. Inertia causes an object at rest to tend to remain at rest and an object in motion to continue to move in a straight line.

Inquisition. A term formally applied to a tribunal (the Holy Office) of the Roman Catholic Church for judgments involving defense of Catholic teaching.

interferometry. A technique that involves using the interference of light waves to measure small angles and from this calculating the diameter of relatively near stars.

interstellar matter. Diffuse gases and microscopic dust particles in the space between stars.

intrinsic (pulsating) variable star. A star that periodically changes in brightness because it is alternately expanding and contracting.

inverse square relationship. A relationship between two quantities such that if one is increased by a certain factor the other is decreased by the square of the factor. For instance, if one quantity is tripled, the other becomes one-ninth as great as it was.

ion. An atom with an electric charge; an atom that has gained or lost one or more electrons.

ionosphere. A region in the upper atmosphere of the earth where many atoms are ionized.

irregular galaxy. A galaxy without symmetrical shape; a galaxy that is neither spiral nor elliptical.

isotope. One of two or more forms of atoms of a particular element that, while having the same atomic number, differ in mass because of differing numbers of neutrons in their nuclei.

Jovian. Of the planet Jupiter.

Julian calendar. A calendar introduced by Julius Caesar in 46 B.C. It fixed the length of the year as 365 days, with 366 days in every fourth year, and it fixed the lengths of the months as they are now.

Kelvin or absolute temperature scale. A thermometer scale in which the divisions are of the same size as those of the Celsius scale; 0 degrees Kelvin (0 degrees absolute) is equal to -273 degrees Celsius.

Kepler's laws. Three statements that describe planetary motion.

Krakatoa. A small island in Indonesia where a violent volcanic eruption occurred in 1883.

lava tube. A natural pipe within a cooling, crusted-over lava flow through which liquid lava flows to feed the advancing flow.

leap year. A calendar year with 366 days.

librations. Motions that in time permit an observer on the earth to see about 59% of the moon's surface rather than just a single hemisphere.

light year. A unit of length equal to the distance light travels in space during one year; approximately $9\frac{1}{2}$ trillion kilometers or 6 trillion miles.

limb. The apparent edge of the sun, moon, or a planet seen against the background of the sky.

limb darkening. The decreased brightness of the sun near its limb as contrasted to the central region of its disk.

line of nodes. The line connecting the two nodes of an orbit.

Local Group. The cluster of about 20 galaxies to which the Milky Way Galaxy belongs.

lunar eclipse. An eclipse of the moon.

magnetic field. The region around a magnetized body within which the magnetism can be detected.

magnetic pole. One of two points on a magnet where the strength of the magnetic field is greatest.

main sequence. The portion of the Hertzsprung-Russell diagram in which most stars are plotted; a narrow band running diagonally from upper left to lower right.

major axis. The longest diameter of an ellipse.

maria (singular: mare). Latin for "seas"; a term used to refer to regions on the moon and Mars that were once thought to be seas.

mascon. A mass concentration beneath the surface of the moon.

mass. A measure of the amount of material in a body.

mean solar time. The interval of time since the mean sun last crossed the observer's meridian.

mean sun. An imaginary sun that moves eastward along the celestial equator (not the ecliptic) at a constant speed, making one circuit in a year.

meridian. A great circle on the celestial sphere passing through the zenith and celestial poles; a great circle on the earth passing through the geographic poles.

meteor. The streak of light produced when a meteoroid enters the earth's atmosphere and is heated through friction.

meteorite. A meteoroid that does not burn completely in its passage through the atmosphere and thus strikes the ground.

meteoroid. A particle or chunk of rock or metal in space.

micrometeorite. A meteoroid so tiny that it moves through the atmosphere too slowly to burn and settles to the ground.

Milky Way. A faint, diffuse band of light encircling the sky and consisting of myriads of stars and nebulae near the plane of our Galaxy.

momentum. A measure of the state of motion of a body; the product of mass times velocity.

moon illusion. A perceptual phenomenon that causes the moon to appear larger when it is near the horizon.

nadir. The point on the celestial sphere directly opposite the observer's zenith.

neap tide. The lowest tides of a lunar month, occurring at first and last quarter when the sun, earth, and moon are arranged in approximately a right angle.

nebular hypothesis. The proposition that the solar system bodies formed from a cloud of gas and dust.

neutrino. A subatomic particle with no charge and essentially no mass; emitted in some nuclear reactions.

neutron. A subatomic particle with no electric charge and mass approximately equal to that of a proton.

neutron star. A small, extremely dense star composed entirely of neutrons.

nodes. The two points where the plane of an orbit intersects another plane, such as the plane of the ecliptic or the plane of the celestial equator.

nova. A star that explosively and temporarily becomes hundreds or thousands of times brighter and thus is seen as a "new" star.

nuclear fission. The breakup of the nucleus of an atom to form two or more lighter nuclei. The process is accompanied by a loss of mass and the release of energy.

nuclear fusion. The formation of heavier atomic nuclei by the union of lighter nuclei. The process is accompanied by a loss of mass and release of energy.

nucleus (of a comet). Rocky, metallic, or frozen material in the head of a comet.

objective. The principal lens or mirror in a telescope; the lens or mirror that gathers light and produces an image.

open (or galactic) cluster. A loose cluster of stars, numbering from dozens to a few thousand, found in the arms or disk of the Milky Way Galaxy.

opposition. The configuration in which a planet is on the opposite side of the sun as viewed from the earth.

optical window. A figurative term referring to the span of the range of wavelengths of light in the electromagnetic spectrum emitted from celestial sources that can penetrate the atmosphere and reach the earth's surface.

optics. The branch of physics that deals with the behavior of light.

orbit. The curved path in which a body moves

around another body as a consequence of gravitational attraction.

parallax. The apparent displacement or shift in position of an object when viewed from different points.

parallax, stellar. The apparent shift in position of a star against the background of more distant stars when viewed from different points in the earth's orbit. Numerically, it is the angle at the star subtended by the radius of the earth's orbit at the star's distance.

parsec. A unit of length equal to the distance at which the radius of the earth's orbit would subtend an angle of one second; the distance at which a star would have a parallax of one second.

penumbra. The partially illuminated portion of a shadow that results when the opaque object casting the shadow blocks only part of the light.

perihelion. The closest point to the sun in a solar orbit.

period. The time required for one revolution of a body around another or for one rotation of a body.

perturbation. A deviation in the orbital motion of a body caused by the gravitational influence of a third body.

phases of the moon. Cyclical changes in the moon's appearance that result when the moon in its orbit presents different portions of its lighted hemisphere to a viewer on the earth. The cycle of phases progresses in this order: new moon, waxing crescent, first quarter, waxing gibbous, full, waning gibbous, last quarter, waning crescent, new.

photon (or quantum). A unit of electromagnetic energy; a discrete portion of energy.

photosphere. The apparent surface of the sun.

planet. Any of nine bodies in orbit about the sun (Mercury, Venus, Earth, Mars, Jupiter, Saturn, Uranus, Neptune, Pluto) or comparable bodies, if there be such, in orbit about other stars. Planets are solid or relatively solid bodies, and they do not emit light.

planetary nebula. A shell of expanding gas surrounding a very hot star.

planetesimal (tidal) hypotheses. Propositions that the planets were formed from one or more masses of material pulled from the sun by a passing star.

population I stars. Young stars associated with interstellar gas and dust.

population II stars. Old stars generally found in globular clusters and galactic nuclei.

precession of the earth. Conical motion of the earth's axis, sweeping out cones in about 26,000 years, caused largely by the gravitational pull of the moon and sun on the earth's equatorial bulge.

precession of the equinoxes. Westward motion of the equinoxes along the ecliptic resulting from the precession of the earth.

Principia. Short for the title of a monumentally important book published by Isaac Newton in 1687; the full title is *Philosophiae Naturalis Principia Mathematica.*

prominence. Flame-shaped or looped masses of gases above the solar chromosphere, best seen on the limb of the sun.

proper motion. The rate of change in direction of a star, measured in seconds of arc per year.

proton. A heavy subatomic particle with a positive electric charge; the nucleus of a hydrogen atom.

proton-proton cycle. A series of nuclear reactions that results in the fusion of hydrogen atoms to form helium atoms with a loss of mass and release of energy. It is the principal mechanism by which the sun emits energy.

protostar. Material in a cloud of gas and dust that is in the process of condensing to become a star.

Ptolemaic system. A concept of the universe in which the earth is at the center and all celestial objects revolve around it, some in very complex orbits.

pulsar. One of a class of objects believed to be dense, rapidly rotating neutron stars emitting bursts of energy in very regular periods of less than a few seconds.

pulsating (intrinsic) variable star. A star that periodically changes in brightness because it is alternately expanding and contracting.

quantum (or photon). A unit of electromagnetic energy; a discrete portion of energy.

quantum mechanics. A highly mathematical branch of physics dealing with atomic phenomena and the emission of radiation.

quasar (quasi-stellar radio source). An object that appears starlike in some of its properties, exhibits a very large red shift, and emits energy copiously.

RR Lyrae variables. A class of pulsating giant stars with periods of less than one day and average absolute magnitudes of about 0.

radial velocity. That portion of an object's velocity that is in the line of sight of the observer.

radiant. A very small area of the sky from which the meteors in a meteor shower appear to originate.

radiation. The process by which energy is transferred through space; also, the energy itself.

radioactive decay (radioactivity). The natural and spontaneous disintegration of an atom. Radiant energy, atomic particles, and a lighter atom result from the disintegration.

radio telescope. An instrument that permits radio emissions from celestial objects to be collected and concentrated by reflectors, amplified, and studied.

radio window. A figurative term that refers to the span of the range of wavelengths of radio waves in the electromagnetic spectrum from celestial sources that can penetrate the earth's atmosphere.

rays. Long bright streaks on the moon. Systems of rays radiate from several craters.

red giant. A large, cool, highly luminous star.

reflecting telescope. A telescope in which the objective is a mirror.

refracting telescope. A telescope in which the objective is a lens.

refraction. The bending of light as it passes from one medium to another medium in which the speed of light is different.

resolving power. A measure of the ability of a telescope to reveal as separate objects two objects that appear to be close together; a measure of the ability of a telescope to reveal detail.

retrograde motion. The apparent westward (backward) motion of a planet with respect to the stars on the celestial sphere.

revolution. The motion of one body around another.

right ascension. The angular distance of a point from the vernal equinox measured eastward along the celestial equator to the hour circle passing through the point.

rille (or rill). A crevasse in the surface of the moon.

rotation. The turning of a body around an axis.

Royal Society. The Royal Society of London for the Improvement of Natural Knowledge; the oldest scientific society in Great Britain, founded in 1660.

saros. A cycle in which similar eclipses of the sun and moon recur at intervals of about 18 years.

satellite. A body that revolves around a larger body; the moon is the satellite of the earth.

Schmidt telescope. A reflecting telescope that uses both a spherical (rather than parabolic) mirror and a thin objective lens. Such a telescope is particularly useful for photographing a wide field of the sky.

second law of thermodynamics. The proposition that holds that in every energy transformation some of the energy becomes heat energy and is no longer available for further transformations.

seeing. An appraisal of the influence of atmospheric disturbances on the sharpness of detail seen with a telescope.

sidereal day. The interval between two successive meridian transits of a star or the vernal equinox; a day as measured according to the stars.

sidereal month. The time required (about $27\frac{1}{2}$ days) for the moon to revolve around the earth and return to its original position against the background of stars.

sidereal time. Star time; the right ascension of the observer's meridian.

sidereal year. The period of revolution of the earth around the sun with respect to the stars; the interval of time between two successive alignments of the sun, earth, and a particular star.

signs of the zodiac. An astrological term for twelve equal segments of the zodiac. Eastward from the vernal equinox, the signs are Aries, Taurus, Gemini, Cancer, Leo, Virgo, Libra, Scorpio, Sagittarius, Capricorn, Aquarius, and Pisces. The signs do not correspond in location to the constellations with the same names.

solar eclipse. An eclipse of the sun.

solar system. The system composed of the sun and all of the objects that revolve about it: planets, satellites, asteroids, comets, and meteoroids.

solar wind. A flow of particles from the sun.

solstice. Either of two points on the ecliptic where the sun is at the maximum distance north or south of the celestial equator.

special theory of relativity. A theory formulated by Albert Einstein that deals with measurements made by two different observers whose frames of reference are in motion at constant velocity with respect to each other. It postulates that the speed of light in a vacuum is the same in all frames of reference.

spectral class (or type). The classification of a star according to the pattern of lines in its spectrum.

spectrogram. A photograph of a spectrum.

spectrograph. An instrument for photographing or recording a spectrum.

spectroheliogram. A photograph of the sun taken in the light of a single element.

spectroheliograph. An instrument for photographing the sun in the light of a single spectral line.

spectroscope. An instrument for viewing a spectrum.

spectroscopic binary star. A binary system that cannot be resolved optically and is detected because periodic shifting of spectral lines reveals orbital motion.

spectroscopy. The science of the analysis of spectra.

spectrum. A band of colors produced when light is dispersed by refraction or diffraction.

speed. Rate of motion or change of position.

spiral galaxy. A relatively flat, rotating galaxy with arms of stars and gas and dust that spiral out from a central nucleus of stars.

spiral nebulae. The original, and inappropriate, term for spiral galaxies.

spring tide. The highest tide of a lunar month, occurring at new moon or full moon when the sun, earth, and moon are in a relatively straight line.

standard time. The time used for convenience throughout a time zone; the mean solar time at a meridian of longitude near the center of the time zone.

star. A sphere of gas that emits radiation as a consequence of nuclear reactions.

steady state theory. A cosmological theory that holds that the universe had no beginning and will have no end, and that matter is being created continually to maintain the average density of matter as the universe expands.

summer solstice. The point on the ecliptic

where the sun is farthest north of the celestial equator. Summer begins when the sun appears to reach the summer solstice.

sunspot. A relatively cooler spot in the sun's photosphere that appears dark by contrast to surrounding hotter regions.

supernova. A stellar explosion in which, within a day or so, a star becomes a million or more times brighter and is nearly destroyed as a star.

synodic month. The time required (about $29\frac{1}{2}$ days) for the moon to revolve around the earth and return to its original position relative to the sun and the earth; the time required for the moon to complete its cycle of phases.

T-Tauri stars. A type of irregularly variable star, believed to be young.

tangential velocity. That portion of an object's velocity that is at right angles to the observer's line of sight.

tidal (planetesimal) hypotheses. Propositions that the planets were formed from one or more masses of material pulled from the sun by a passing star.

tide. The deformation of a body of land or water when the gravitational forces from another body pull unequally upon different portions of the body.

transit. The passage of a planet across the disk of the sun or of a celestial body across the meridian.

tropical year. The period of revolution of the earth around the sun as measured by the seasons; the interval of time between two successive passages of the sun through the vernal equinox.

umbra. The completely dark, central portion of a shadow; also, the dark, central portion of a sunspot.

velocity. Speed in a specific direction.

vernal equinox. The point where the sun appears to cross the celestial equator moving northward. Spring begins when the sun appears to reach the vernal equinox.

visual binary star. A binary system in which two stars can be resolved (that is, seen separately) with a telescope.

walled plains. Very broad, shallow craters on the moon.

wavelength. The distance between corresponding points on adjacent waves.

weight. A measure of the gravitational force between the earth and an object on or near its surface, or between other celestial bodies and objects on or near their surfaces.

white dwarf. A small, hot, dense, faint star believed to be near the end of its evolutionary development.

winter solstice. The point on the ecliptic where the sun is farthest south of the celestial equator. Winter begins when the sun appears to reach the winter solstice.

zenith. The point on the celestial sphere directly above the observer.

zodiac. A band on the celestial sphere about 16 degrees wide, centered on the ecliptic.

Index